'TRAX 3'

SIGNALLING
LEVER FRAMES

Jeff Geary

CW00516662

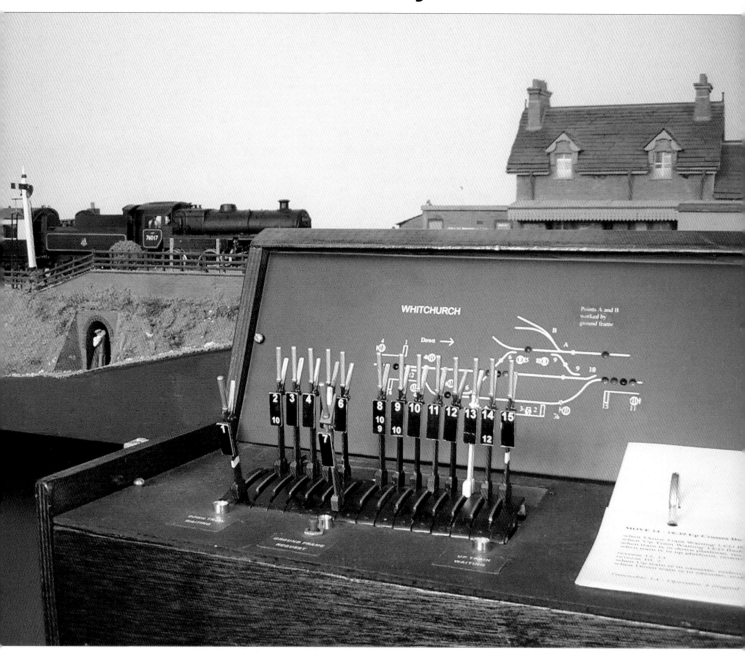

Noodle Books

© Jeff Geary and Noodle Books 2011

CD ROM © Jeff Geary 2011

ISBN   978-1-906419-61-5

First published in 2011 by Kevin Robertson
under the **NOODLE BOOKS** imprint
PO Box 279
Corhampton
SOUTHAMPTON
SO32 3ZX

**The Publisher and Author hereby give notice
that all rights to this work are reserved.
Aside from brief passages for the purpose of review, no
part of this work may be reproduced, copied by
electronic or other means, or otherwise stored in any
information storage and retrieval system without written
permission from the Publisher. This includes the
illustrations herein which shall remain the copyright of
the copyright holder.**

www.noodlebooks.co.uk

Printed in England by Ian Allen Printing Ltd.

(All unaccredited images are either by the Author or from the Noodle Books collection)

CONTENTS

INTRODUCTION

As a small boy I, like millions of others, would go train-spotting. One of the joys of that pastime was waiting by the lineside fence and watching the signals. Whether it was the satisfying 'thump' of a semaphore arm coming off, or the silent substitution of a green for a red light, a sense of anticipation was created. The thrill would mount as the rails began to hiss and then finally there would be the engine itself thundering towards us. Eyes glued to the cabside, we would spin our heads quickly to follow its number as the engine flashed by and then check our Ian Allen book to see if, joy of joys, it was a number that hadn't yet been underlined! Finally, as the train disappeared into the distance, like the curtain coming down on a small act of a drama, the signal would return to danger and we would begin our long wait until the next couple of minutes of excitement.

Since I was persuaded, mostly by the example of my friend John Shaw, that model railways should be signalled properly, I have come to realize that this excitement can be created in miniature. If we take our club layout to an exhibition, the same sense of anticipation can be created by telling a small child "watch that signal". When the arm drops (or rises) to signal the approach of a train, hushed expectation is followed by squeals of excitement as the train comes into view. For the operators, there is a different kind of satisfaction. To work with your fellows through that complicated choreography of train movements that is a working timetable without a single word being exchanged, all done by careful attention to the signals, gives a tremendous sense of achievement. This is not playing trains, this is operating a railway in miniature!

This book is designed to help you create a properly signalled railway and to build a replica of a signalman's lever frame, or a more up-to-date control panel, to operate its points and signals. Where to position the signals, how to assign them on the lever frame, how to use locking to ensure that conflicting signals and routes cannot be set, and all the other constraints that must guide the design of your model are covered. As with our earlier books *Wiring the Layout* and *Track Construction*, we have included with this book a free CD containing the latest release of my *Trax* program – *Trax 3*. This has the same capabilities as earlier versions of *Trax*, plus a whole range of new facilities. I hope you will find it useful

Jeff Geary

ACKNOWLEDGEMENTS

As ever, I would like to thank Kevin Robertson and John Shaw for their help and advice. John also produced most of the track diagrams in chapters 3 and 4. For their kind permission to use their photographs I should also like to thank Dave Renshaw (fig. 1.3), John Tilly (figs 2.4, 2.8, 2.9, 3.1), Graham Hatton (fig. 2.5), Dr Neil Clifton (fig. 2.10), David A. Ingham (figs 2.12, 3.7, 3.8 and plate I), John Saxton (fig. 2.13), and Geof Sheppard (plate II). Many of these I located on the world wide web, and the reader is encouraged to look at the websites listed in Appendix 1 to see more of their work, and to find extensive and detailed information on this fascinating subject.

From the sublime to the.....Mechanical semaphore signals operated by differing means. On the left are electrically operated signals at Winklebury (west of Basingstoke), for the Up Main, Up Relief, and Up Main to Up Goods Loop. The respective signal motors are also visible. On the right is the station at Cranleigh (between Horsham and Guildford) looking towards Baynards. Here the starting signal has a Shunt Ahead signal on the same post with a second Ground Signal in the 'six-foot' covering wrong-road movements from the opposite platform. The two subsidiary signals were of the 'selected' type and meaning the locking allowed the same lever to operate either dependent upon the position of the associated pointwork.

Interior of Taunton East signal box, April 1985. Whilst the box dates from GWR days there is a mixture of original and later BR equipment. In the former category are the brass lever leads: the ivorine type first seen around the time of WW2 and was later used extensively whenever replacements were required. The sequence of levers in the frame generally mirrored the position of the associated items on the ground, although down line signals were at one end of the frame and up line signals levers at the opposite end. In a box such as this with in excess of 140 levers, that could mean an awful lot of walking on the occasions when traffic was slack and just one man might be on duty - usually night time. Notice the two detonator placing levers. Both are free to be operated at any time (not interlocked with anything), those seen are identified as being for the down line, as the painted chevrons on the levers face downwards.

1

DIFFERENT WAYS TO RUN A RAILWAY

In this chapter, various methods of controlling model trains are reviewed, and the personal preferences of the author will become apparent. Modellers vary in the importance they attach to different aspects of the hobby, and this is one of its appealing features. Your interest may be in building locomotives, or in scenic modelling, or electronics, or any other of the many aspects. This book focuses on operating the model railway, and suggests that the most satisfying way to do this is to simulate, in miniature, the way a 'real' railway works. That is, trains are run to a working timetable and are controlled by a realistic network of signals, and these signals are best operated by a proper lever frame. This can be made just as important a part of the model as the trains themselves and it can be placed out in front of the layout rather than tucked away out of sight. Later in the book, we shall see how an interlocking lever frame can be built, how it can be interfaced to the points and signals on the layout, and to other lever frames if necessary, and how the operation of the railway can be made as lifelike as possible. We start by briefly reviewing the development of different methods of controlling model trains.

A Short History of Train Control

At the most basic level, trains can be controlled entirely manually. This is not uncommon among the fraternity that collect and run old tin-plate clockwork models. To make a train go, the engine's spring is wound up with a key and the brake lever is released. The engine will then haul the train around the track at maximum speed, gradually slowing as the spring unwinds. To stop the train before the spring runs down completely, you grab hold of the engine and put the brake lever into position to lock the clockwork mechanism. Routes are set entirely manually, changing points by means of a lever adjacent to the switch end of each point. Although such models have a certain charm, and are historically quite interesting as examples of toys from a bygone age, this method of operation bears very little resemblance to the way a real railway works. Over the years, more sophisticated schemes have evolved to allow model railways to be operated in a more realistic fashion.

This has been made possible largely as a result of the replacement of clockwork by electrical operation. Instead of a manually wound spring, the power to drive the engine is derived from electrical current fed to it through the track. In its earliest form, alternating current (AC) would be used, allowing a step-down transformer to be connected to the domestic mains supply on its primary side, and to the track on its secondary side (though see figure 1.1). This current was picked up through the engine's wheels and (in some cases) a brass plate making a sliding contact. A large

Figure 1.1

An electric train set without an electricity supply. An early American set powered by a hand-operated generator.

7

rheostat would allow the track voltage to be varied, thus enabling the operator to control the speed of the engine from standstill to maximum speed and back again.

DC Control

With the development of cheap and efficient rectifiers, direct current (DC) gradually superseded AC, and this had the further advantage that a DC motor can be made to run either way simply by altering the polarity of the applied voltage. Thus by adjusting the track voltage and polarity, the speed and direction of the engine could be controlled remotely. This was certainly a big advance in the realistic appearance of model railways. The ability to stop a train without a giant hand descending from above and grabbing hold of it made a big difference, as did the ability of a train to start slowly and build up speed gradually. Shunting with an engine (as against with the afore-mentioned giant hand) became possible. Further developments included the introduction of solenoids and or motors to operate points remotely. These enable a route to be set by means of a series of toggle switches or push-buttons, often located on a control panel, upon which an outline track diagram is depicted providing an indication as to which switch does what.

Further enhancements to the control panel concept included the introduction of section switches, which would allow different sections of track to be switched in or out, and the introduction of multiple controllers, so that several trains could be operated independently. With ever larger and more complex layouts the concept of *cab control* was developed, whereby several operators could each 'drive' their own engine through the layout by virtue of toggle switches connecting their controller to a series of track sections. More sophisticated cab control schemes would switch track sections automatically, using relays, as trains progressed through the layout. Larger layouts would incorporate multiple control panels requiring many operators to run a complex and busy schedule of trains.

All this, however, came at a price. Because of the scale and sophistication of some layouts, the task of wiring and maintaining 'the electrics' became enormous. Every length of track that might be required to be isolated, every point, every working signal, turntable, and lineside feature would have to be individually wired back to the control panel where its operating switch or potentiometer was located. If the layout was to be transportable, to take to an exhibition or a club meeting for example, then this wiring would have to be dismantled and re-assembled (unless the layout was a very small one that fitted on a single baseboard). This would involve multi-pole connectors and cables, adding further to the hundreds, or maybe even thousands, of solder joints already existing on the layout. This just increased further the chances of a failed solder joint, a short-circuited

wire, a damaged connector pin, and all the other potential sources of problems so often encountered with 'the electrics'.

Digital Command Control

It was in part to alleviate this explosion in wiring that the concept of *Digital Command Control* (DCC) was developed. There had been some similar systems around since the 1970s, notably in this country the Zero 1 system from Hornby. However these were specific to one particular manufacturer's locomotives, and such locomotives could not be mixed with 'normal' DC-operated locomotives on the same layout. It was not until 1994, when the National Model Railroad Association (NMRA) in the USA issued its Standards for Digital Command Control, that DCC started to become widely used. The standards ensure, among other things, that products from different manufacturers can be used together on the same layout.

In a nutshell, DCC makes all of the track 'live' all of the time. Inside each locomotive is a micro-computer chip, referred to as a decoder, which can be given individual instructions from a command station to start up the motor, slow it down, reverse direction etc. All locomotives receive all of these instructions, but a given locomotive will only respond to those instructions specifically addressed to itself by way of an identifying number. In this way, the whole layout can be wired as a single entity, and all the complexity of track sections, isolating switches, etc. can be confined to the dustbin. Furthermore, the concept can be extended to

Figure 1.2 - A Wiring Explosion!

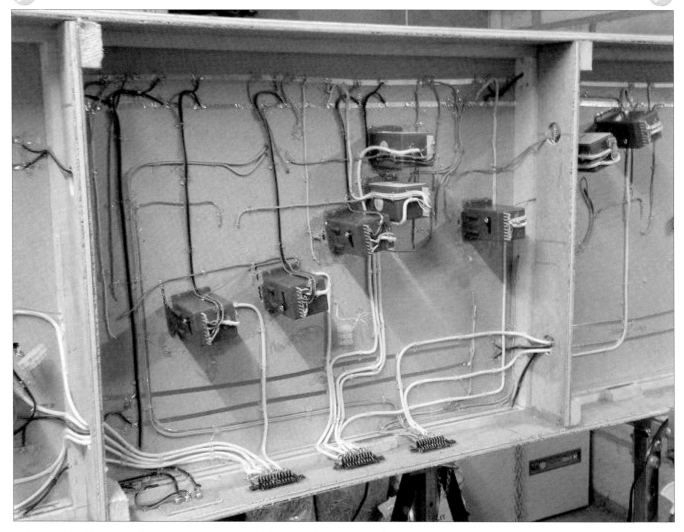

Figure 1.3 - *Underneath Dave Renshaw's DCC layout 'Cramdin Yard'.*

equipment other than locomotives. Points and signals, for example, can be addressed by identifying numbers and instructed through the medium of the track in the same way as locomotives. The user communicates these instructions from a small keypad, of which there may be more than one, thus allowing several operators to drive trains independently on what is, from an electrical point of view, the same track. The potential for reducing the complexity of track wiring is immediately obvious – compare figures 1.2 and 1.3.

DCC can offer other benefits, too. You can program the command station to recognize your locomotives individually, and tell it, in effect, "this one's a bit of a laggard and needs lots of power to get it going" or "this one's got a really fast motor – keep the volts down". You can program in inertia and braking, or you can get two locomotives to go at exactly the same speed – useful for double-heading or banking. The presence of continuous power at all points on the track means that realistic lighting and sound effects can be achieved. Passenger coaches can be illuminated, even when standing in the platform, engines

and brake vans can show head and tail lamps, you can have a light in the firebox (even one that brightens up as the locomotive is called upon to work harder!). Some of the sound effects chips are very impressive: Not only can you have an exhaust beat when the engine is in motion, but you can have safety valves blowing off, steam whistles, and a host of other sounds. All in all, DCC has a lot to offer.

My first 'hands-on' experience of DCC came some time ago when I had the privilege of operating a very fine layout representing a logging line in 0 scale. The scenic modelling was to the very highest standard, and maximum use had been made of DCC facilities to make the operation of the railway breath-takingly realistic. Bells would clang when locomotives reversed, brakes would squeal when they slowed, escaping steam would hiss, and so on. It really was very impressive indeed, and I told the owner what I thought. He thanked me, and went on to remark how many dinosaurs there were in the world. "DCC is the only way to go in the future" he said, and everyone who was stuck in the old DC wiring world was a dinosaur! I had been acquainted with

Figure 1.4 - Inside Blue Anchor signal box on the West Somerset Railway.

in the time period I model!). I had another first 'hands-on' experience when I first operated John Shaw's Aveton Gifford, a layout mentioned in my earlier book *Wiring the Layout*. This was the first time I'd used a lever frame with full mechanical interlocking to control all the points and signals, and it was fascinating. Since that experience I've rebuilt my own layout (twice!) to use the same principles, and helped rebuild parts of our local club layout, which now boasts five lever frames.

"So what's so special about lever frames?" you may ask. The answer, quite simply, is that it was (and still is, in some cases) how real railways worked. There is a great deal of satisfaction to be had from working out, from a track diagram, what lever pulls must be made to signal a given move. It is not always easy, but there is a great sense of achievement to be had from learning, and eventually mastering, a lever frame. Once you get to a large layout, where several operators need to combine their lever pulls to make things work, there is a splendid camaradie that develops when things go well (and vice versa, but it's still fun!).

After the footplate of a steam engine, the inside of a signal box is probably the most evocative place on a railway. The hand-painted track diagram, the polished wood and brass of the indicating equipment, the shiny linoleum floor and the lever frame with its brightly painted levers - red or yellow for signals, black for points, blue for locks – all combine with the smell of paraffin lamps and metal polish to evoke a wonderful atmosphere. This, to my mind, is how railways should be!

There are kits around to enable you to build your own twelfth scale lever frame, complete with full mechanical interlocking exactly like the prototype. (This use of the word 'prototype' always seems a little strange. In the engineering world a prototype is a working model of what will eventually become the production item. Its use in railway modelling, to mean the 'real' railway upon which the model is based seems somehow the wrong way round! However, years of usage have meant that it is now established.) The main thesis of this book is that a model railway can be operated entirely through a prototypical lever frame and that, with the possible exception of a direction switch on the locomotive controller, there should be *no switches*. Everything is worked through the lever frame.

DCC and Lever Frames

The question arises, whether you can use DCC alongside a prototypical lever frame. The short answer is yes. However, there are a number of issues that you might need to consider before deciding whether and how to combine the two. To see what these are, we need to have a very brief look at how a real railway is controlled. We will look at this

this chap for some time and had been well impressed, inspired even, by some of his previous layouts. I therefore had a great respect for his opinions. Why, then, have I not been totally converted? Why am I still a dinosaur? Indeed, why are there (as had been remarked) still so many dinosaurs in the world?

The Lure of Lever Frames

A common reason for not converting to DCC is the cost. If you happen to have a large stock of older locomotives then the cost, both in money and modelling time, of conversion becomes considerable. If your models use coreless motors this can mean you need special decoders for them. However, cost was not the sole issue for me – I had already been lured away by a completely different method of operation

Although DCC makes for a model railway that is wonderful to look at and listen to, operating it is somewhat akin to programming your TV to do a timer recording using the remote. Proper trains don't work through little keypads (well I guess some of them might nowadays, but they didn't

in more detail later, but for the moment some very general remarks will suffice.

On the prototype, a railway line is divided up into what are called block sections. A train can only enter a section when the stop signal controlling that section is pulled 'off' in the lever frame. The train can then proceed to the next stop signal, at which it must halt, unless that signal too has been pulled off, in which case it can proceed to the next block section. At junctions and the like, different signals control access to the different routes through the junction, and interlocking in the frame ensures that a given signal can only be pulled off once the points have been correctly set for the chosen route.

Now, you may spot the fact that this method of working bears a lot of similarity to the cab control schemes briefly described earlier. If you have a layout with a junction between a main line and a branch, and signal it appropriately, then the lever that signals a move through the junction along the main line could very easily also operate a relay to connect that block section to the main controller. Likewise, the lever signalling a train onto the branch line could operate a relay that connects this section to the branch controller.

This is a very simple and natural way to wire things up. However, it does mean plenty of wire and connectors with all the attendant maintenance issues described above. To do it the strict DCC way, I suppose you could dispense with this wiring and use the levers to operate some electronics to send suitably encoded instructions to the appropriate point or signal through the track.

However this does seem a roundabout way of doing things. It would be simpler to wire your points and signals up directly to the levers. There then remains the question of what to do about the block sections. You could, of course, ignore these and wire the layout up 'all live' as in standard DCC, but having gone to all the trouble of ensuring, by interlocking in the frame, that conflicting signals and route settings cannot occur, do you really want to give your train drivers the ability to drive through signals set at danger and stage head-on collisions (or 'corn-field meets' as the Americans quaintly call them)? You may feel it is safer to arrange matters so that an engine is forced to stop at a stop signal (or shortly after it) by virtue of power being removed from the track.

For these reasons, some of the advantages of DCC would not be applicable to a lever frame based scheme. You would still need a lot of wiring, so the claim of DCC proponents that wiring is reduced to a minimum would not apply. Nevertheless, there is no reason why, in principle, a DCC signal should not be used to control your locomotives, even if the points and signals are operated from the frame.

The advantages of realistic inertia, braking and sound would all still be available. If you wish, you can wire the whole layout 'live' as per standard DCC, or you can switch the DCC power according to signal settings as described above. I have no direct experience of this, but cannot see why it wouldn't work.

So, although the short answer to the question of whether DCC and lever frame control are compatible was 'yes', you might want to consider just how much use you make of DCC and how much you use conventional wiring from the lever frame.

About This Book

In the remainder of the book, we shall review (briefly, for it is a vast subject) the development of signalling on the prototype. The emphasis here will be on the principles of signalling and why it evolved in the way it did.

In the next chapter (3), we shall look in more detail at the different types of signal and their significance. We will then go on to look at some typical track layouts and how they are signalled on the prototype.

In chapter 4, we look at adapting prototype signalling practice to the model layout, and we see how the lever frame can be used to provide total control of the railway. The key here is *no switches*, everything works through the lever frame.

In the following chapter (5), we shall discuss the construction of lever frames, showing how a locking frame can be built incorporating all the locks and releases required. The process of developing the actual mechanical layout of the locking bars from the theoretical locking chart will be discussed in detail.

An alternative to mechanical locking is the subject of the next chapter (6), namely electrical locking by means of relays. On the prototype, as colour-light signals replaced semaphore arms, so relay locking replaced the mechanical locking frame. We describe how all of the necessary locks and releases can be achieved by using relays.

The final two chapters will introduce the CD accompanying this book which includes the third version of my *Trax* program, *Trax3*. Once again, it should be emphasised that *Trax* is offered 'as is'. Whilst care has been taken over testing the program, and it has been checked on a number of operating systems, no guarantees are offered that it will do exactly what you want. If you do find what you think might be a bug, then by all means report it via the publishers, and if it can be readily fixed it will be.

New facilities available in *Trax3* include some new track formations such as scissors crossings and transition curves, the inclusion of relays on the track diagram to help in checking your relay logic and extended drawing facilities. As you might expect (given the title of this book) several signal-related features have been added: new signal types and lever frames now incorporating full locking facilities. Also added is a scripting facility to develop and test your timetabling.

In addition to *Trax3*, the CD includes a couple of small programs you may find useful in developing and checking out your locking charts. *FrameTest* produces a list of all the lever combinations your locking rules allow. You can then check these out on *Trax3* to see if your locking is complete and safe – i.e. that all the combinations it permits are valid ones and none result in shortcircuits, etc. *LockingBars* is a little program which helps with the tricky business of working out, from a locking chart, a mechanical design for a set of locking bars.

Figure 1.5 - Signalling in one-twelfth scale. John Shaw built this lever frame for Phil Abbott's model of Merthyr Tydfil. Notice also Phil's miniature block instruments.

2

SIGNALLING PRINCIPLES

In this chapter, we shall review how and why railway signalling developed. The subject is a vast one, with a great many books written about it, a selection of which are listed in the Appendix 1. What follows is, of necessity, very brief and highly selective. Its purpose is to give the reader an idea of the general principles, and the reasons for them, so that we can apply the same principles to signalling a model railway and thereby ensure it is as authentic as possible.

In the beginning, trains were controlled by hand signals, given by what was then called a policeman. The Liverpool and Manchester Railway (1830) employed an establishment of around 60 policemen to signal the 31 miles of track. If a policeman held his arms outstretched, this signified 'line clear' and the engine driver could proceed. If he stood at ease (or could not be seen), this signified 'danger' and the engine driver was to proceed with extreme caution. Other railways had different conventions, however. On the Great Western Railway, an arm held out horizontally signalled 'line clear', upright it meant 'caution', and two arms held upright signalled 'danger, stop'. Whether or not the line ahead was clear was determined by the period of time that had elapsed since the last train passed, and each policeman had a watch or sand timer to measure this. At night, hand signals were obviously not useful, so the policeman was equipped with a lamp having differently coloured glasses so that he could show a red light for 'danger', green one for 'caution' and white for 'clear'.

One problem with hand signalling was that it required the policeman to be more or less permanently in position. Before long, therefore, mechanical signalling was developed, so that a policeman could set a signal then go off

and attend to other duties. There followed a great variety of signalling devices, ranging from flags of different colours hoisted up masts to wooden boards of various shapes and colours pivoted to turn and show different aspects. The Great Western Railway erected a tall pole at Reading station, up which a red ball was hoisted to signal 'All Clear'. The ball was lowered to the ground to signal 'danger'. Daniel Gooch famously said of this signal 'If the ball is not visible, the train must not pass it'! A particularly spectacular signal was devised by the Eastern Counties Railway which had five coloured panels, two red, a green, a yellow and a white. The panels were each shaped like a segment of pie and arranged to slide behind each other rather like a Victorian lady's fan. All five were displayed immediately after a train had passed and this indicated 'Danger – Stop'. After five minutes the lower of the red panels was concealed, indicating a slightly reduced level of danger. At two and a half minute intervals thereafter, the remaining panels were concealed, until only the white panel was visible, indicating 'All Clear'.

Semaphore Signals

However, all of these contraptions were eventually superseded by the semaphore signal, first used in a railway context by Charles Hutton Gregory on the London and Croydon Railway in 1841, although similar devices had been used by the Royal Navy for signalling since the 18th century. The semaphore signal employed a wooden arm mounted on a pivot at the top of a tall post. The arm extended horizontally to indicate 'danger'. By means of a lever at the base of the post, the arm could be moved so that it inclined downwards at 45° to indicate 'caution', and yet further so that it hung down vertically in a slot in the post to signal 'clear'. Adoption of this simple but effective signal was rapid, and by 1860, most railways had adopted it. At night, a lamp was mounted on the signal post displaying a red, green, or white light in the same way as the policeman's hand-held lamps referred to above. Often, two arms were mounted on the top of a single post, one for each direction. When seen from the engine driver's viewpoint the signal that applied to him was always on the left as he approached the signal (figure 2.2).

These early semaphore signals would be operated as the mechanical equivalent of the policeman's hand signals. That is, after a train had passed it, the signal would be put to 'danger'. After five minutes, the arm would be moved to the 'caution' position, then after a further five minutes it would be moved to 'clear'. The engine driver on a second

Figure 2.1 – Early hand signals (GWR)

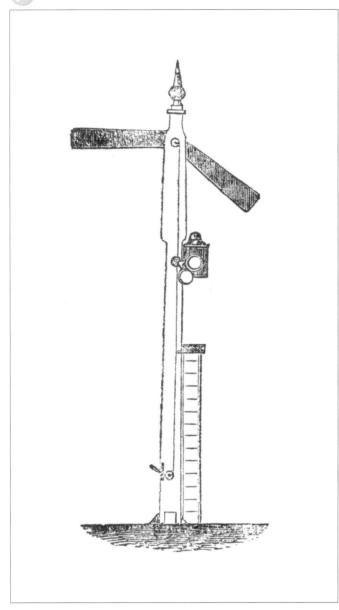

***Figure 2.2** – An early semaphore by Stevens & Co.*

train on the same line would, upon seeing the 'danger' signal, bring his engine to a halt and wait for the 'caution' indication, whereupon he could proceed at a reduced speed. If the signal were at 'clear', then he could proceed at full pace. The five minute intervals were not universal – on some lines they might be more, or less, depending on conditions.

The idea of this 'time interval' signalling was to keep trains at least five minutes apart. If both trains ran to time, it worked. However, if one train encountered a problem such as a mechanical failure, which brought it to a halt, then there was considerable danger. What was supposed to happen was that the fireman or guard would run back down the line so that he could bring a following train to a halt by means of

hand signals. However, train braking at that time was not good, perhaps just a couple of handbrakes on the tender and the guard's coach. The risk of a collision in such circumstances was therefore considerable.

As traffic built up on the railways, the time interval method of signalling would mean that eventually all trains would run at the speed of the slowest. At intervals along the line, therefore, sidings or loops would be provided so that slow goods trains could be moved out of the way of faster passenger trains. Whilst such manoeuvres were under way, points would obviously have to be set for the sidings, and signals would have to be employed to prevent trains on the main line from running over them whilst they were so set.

Operation of Points

Points, like signals, would be operated in situ in the early days of railways. Each point would have a lever alongside working the switch blades directly. The process of signalling a train through a station could therefore involve much running around, first to set points for the correct route, and then to set the appropriate signals. The system was also obviously exceedingly dangerous, in that it relied on the staff to remember which points were set which way.

The following anecdote from Hubert A. Simmons, writing in the nineteenth century under the pseudonym 'Ernest Struggles' of his experiences in 'The Life of a Station Master', illustrates why. Ernest encounters Porter Broom, who has the duty of changing points and signals, the station having no signalman:

'I never saw Broom without refreshing his memory, and, from habit I suppose, I said to him, "Have you put that signal right for the express?"
"Yes, sir" replied Broom.
"And did you put the points right and lock them?" I further enquired.
"They were all right this morning", replied Broom.
Not a moment's hesitation, not a second thought or enquiry, I did not even stop to close my cash drawer, but I knocked a passenger over who was stood in the way of my shoulder, and I was running, not for my life, but for the lives of others, to the points, for I knew that if Broom had not put them right since the morning they must be wrong. The signal was indeed 'all right' but the points were 'all wrong', and the safety of the unconscious load of passengers depended on my turning them before the express train got there. She was coming 'full steam' and I was going 'full steam'. Broom, too, was coming behind me, but I had distanced him, for I could run.
If you had been an observer of the race I think you would have decided against my chance of success, but the excitement of the moment lent me the wings of lightning, as it were.

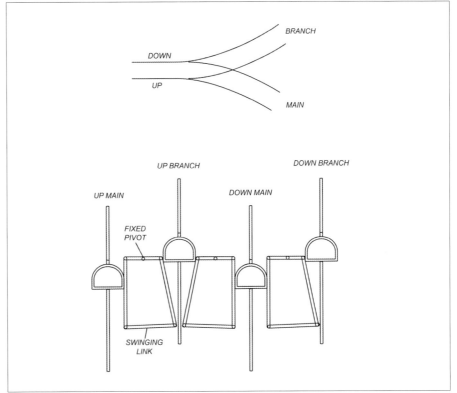

Figure 2.3 – *Gregory's stirrup frame with up and down main signals set 'clear'. Other valid combinations would be up and down branch, or up main and down branch.*

The train was coming like a charge of cavalry, the points were between us. I had to stand on the line where the train would pass if I missed the handle. It was death or victory.

Another moment and I should have been a mangled corpse or the train would have passed the station in safety.

I need not tell you that it was the latter, I had turned the points and had staggered back into a muddy ditch from which Broom extricated me, for my strength had gone.'

Although amusing, this story draws attention to the very serious problems that could occur when the settings of points and signals were conflicting. A great improvement came with the grouping together of signal and point operating levers. These would be mechanically connected to the signal posts and points by means of wires (in the case of signals) or rigid steel rods (for points). Having the levers grouped together meant that the operator could see at a glance which way they were set. Gradually, the term 'signalman' came to be applied to this person, replacing the word 'policeman'. Colloquially, however, for ever afterwards signalmen were known as 'Bobby'.

Interlocking of Points and Signals

Charles Hutton Gregory installed a device in 1843 at the Bricklayers Arms junction of the London and Croydon and South Eastern Railways whereby conflicting signals could not be given (Figure 2.3). The signals were operated by foot pedals, rather like stirrups, which when pressed down moved either one or two swinging linkages. These then obstructed the adjoining stirrups which might otherwise set conflicting signals. His system did not, however, preclude points being set wrongly, since these were worked by separate levers alongside the stirrups.

In 1856 John Saxby of the London, Brighton and South Coast Railway created a forerunner of the locking lever frame. This had levers that set the points arranged so that they operated the signals simultaneously, so that a signal was only set to 'clear' when its corresponding route was set by the points. In 1861, Saxby set up his own firm to manufacture signalling apparatus, and the following year, with John Farmer, founded the firm of Saxby and Farmer, which became a major supplier to many railways.

However, the first interlocking frame in the modern sense was the work of Austin Chambers of the North London Railway, who improved a lever/stirrup frame supplied by Stevens and Sons, rivals to Saxby and Farmer. He devised a mechanism whereby the points levers moved metal plates fitted with holes through which rods driven by the signal stirrups passed. Until the holes in the plates were correctly aligned, the signals could not be operated. Thus by ensuring that the holes were lined up for a given signal rod only when points were set correctly Chambers achieved the desired result. The apparatus was installed at Kentish Town in 1860.

Various other schemes for interlocking appeared around this time, however the one that became almost universal was

Figure 2.4 *– Tappet locking bars*

tappet locking, pioneered by Stevens and Sons. We shall look in detail at this form of locking in subsequent chapters, for this is by far the easiest scheme to implement mechanical locking in a small scale model. In figure 2.4, the point and signal levers on the extreme left move sliding bars called *tappets*, which run left to right in the picture. If the lever moves, the tappet slides. If the tappet is prevented from sliding, then the lever cannot be moved. By means of specially shaped blocks called *dogs* which are pushed by cut-outs in the tappets, *locking bars* can enable or disable certain levers, depending on what levers have been pulled already. The locking bars are the square-section bars running from front to back of the picture. Details of the scheme will appear in chapter 5.

The safety benefits of such a system are clear. We can arrange that point levers release signal levers, such that a signal lever can only be moved, putting its signal to the clear position, if all necessary points have been set correctly. Once a signal lever is pulled, it can then in turn lock any other levers that might signal a conflicting movement, as well as locking the point levers in position.

Facing Point Locks

A facing point is one which presents to an approaching train its 'toe' end first (the toe is where the switch blades are located). There are some special considerations that must be taken into account with facing points, and in general they are to be avoided where possible. A trailing point is one which is normally approached by a train at its 'heel' end. This means that the train could conceivably run into a trailing point the

'wrong way', although interlocking of points and signals should prevent this. Even so, this would be less likely to damage the train than to damage the points. Of course, there are situations where facing points are inevitable, for example at junctions and on single lines where there is a passing loop.

At junctions, the 'straight on' line will often have a higher speed limit than the diverging one. If a train is approaching the junction, it is essential that the driver is aware of the setting of the point, since if he approaches it at high speed when it is set for the diverging line, the train could de-rail.

Figure 2.5 *– A facing point lock with detector bar just inside the running rail. One of the slots in the stretcher can be seen left of the plunger.*

This can be indicated by means of a pair of semaphore signals, one for each line; however to further ensure that the correct line is signalled a facing point lock can be employed.

A facing point lock comprises a stretcher bar that fits between the two switch blades at the toe end of the point, in which there are two slots. A plunger can pass through one or other of these slots when the point is set fully left or right. Once the plunger is in the slot, the point is locked: it cannot be moved again until the plunger is withdrawn. In the lever frame, the point is operated by one lever, and an adjacent lever operates the locking plunger. Only when both the point lever is set, and the locking lever is in the 'lock' position can the appropriate signal lever be pulled.

Facing point locks are often also provided with detector bars. These are long steel bars, operated by the same lever as the locking plunger, and run parallel to and just inside one of the running rails. Whereas the bolt is moved in a straight line, into or out of the slot in the locking stretcher bar, the detector bar is moved on a pair of short cranks which travel in an arc. Thus the detector bar, as it moves forwards and backwards also rises up to rail height. If there happens to be a train standing on the running rail, the detector bar will be prevented from lifting by the wheel flanges. This ensures that the signalman cannot change the points whilst a train is standing over them.

The Electric Telegraph

The improvements in safety arising from the development of the locking lever frame were considerable. However, it did nothing to improve the lamentable situation referred to earlier, that the decision whether to assume the line ahead was clear was based solely on the length of time that had elapsed since the last train passed. This was clearly a very unsatisfactory scheme.

The invention of the electric telegraph changed all this. In 1837, William Cooke and Charles Wheatstone installed a demonstration at the Euston headquarters of the London & Birmingham Railway. Operation of a key at Euston caused the flick of a needle a mile away at Camden Town. However, the L&BR did not develop the potential of this equipment, and it was removed after a few months. In 1839, the GWR installed the electric telegraph between Paddington and West Drayton, a distance of 13 miles. It was not, however, used for signalling, but for general communication purposes.

In 1842, Cooke wrote a pamphlet entitled 'Telegraphic Railways', in which he recommended that lines be divided up into sections, and that only one train at a time should be allowed in each section. Telegraph instruments should communicate the entry and exit of trains to sections ahead and behind, thus allowing signalling to be based not on a

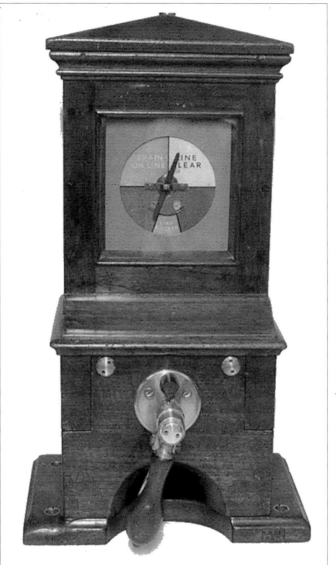

Figure 2.6 – An LNER Block Telegraph Instrument.

time interval, but on physical separation of the trains. The first railway to be so signalled was the Yarmouth & Norwich Railway, which was equipped in 1844 with Cooke and Wheatstone's Railway Telegraph at each of the five stations along its 20 mile length.

Edward Tyer's version of the railway signal telegraph machine was patented in 1852, and first used on the South Eastern Railway. This employed two needles to indicate the state of the line in rear and advance of the apparatus. (To explain: if you were to stand on the line, facing in the normal direction of travel, then everything in front of you is 'in advance' and everything behind you is 'in rear'). Each needle could point to either 'Blocked' or 'Clear', and unlike earlier machines it remained so until set otherwise. As a train passed from the section of line in rear of the apparatus, to the one in advance, the operator would set 'Line Blocked' for the section in advance, and 'Line Clear' for the section in rear. These indications would be repeated at the adjacent boxes, whose operators would set similar indications for their sections. In addition, the instrument had a bell that could be rung by the an operator at an adjacent station to call attention, or to indicate the type of train that was on its way.

Tyer's machine (and others similar to it) offered obvious safety benefits. However, it was not until 1889 that it became compulsory to use block section working. In June of that year, an appalling collision occurred in Armagh between a regular service train and coaches full of school children and their parents on a Sunday school trip. The casualties, 78 dead and 260 injured, many of them children, caused such an outcry that within six months the Regulation of Railways Act was passed, making the use of block section working compulsory on all passenger-carrying lines, along with interlocking of points and signals, and the fitting of continuous brakes to all vehicles on a passenger-carrying train. These and numerous other regulations came to define how railways were to be operated.

Block Section Working

Since it is the key to the safe operation of a railway, let us look at the general principles of block section working, as it came to be practised after the 1889 Act. Note that this a brief description, not a fully detailed account, which would fill a book in itself. All that is required for our purposes is an idea of what is involved so that we can base our model upon the same principles.

We will suppose that there are two adjacent signal boxes, A and B. At each signal box, there are two stop signals for each direction, one in rear of the signal box, referred to as the home signal, and one in advance of the box, referred to as the starter signal. Considering the case of an 'up' train, the block section is that length of line which lies between A's up starter and B's up home signal. The line between each of their home and starter signals is referred to as 'station limits'. The essence of block section working is that the block section between A and B in the up direction can only be entered from A with B's permission, and B will only give this permission when he is sure that the line between them is clear.

The sequence of events is as follows. A train is standing at A's 'up' starter waiting to go to B. The signalman at A will first of all call B's attention by a single ring on his bell. B replies by a single ring, indicating that A now has B's attention. A will then send a sequence of rings called a 'bell code' which serves a dual purpose. Firstly, it indicates the type of train, for example four consecutive beats meant an express passenger train, three beats-pause-one beat an ordinary passenger train and so on. A list of some common bell codes is given in appendix 2. The second purpose served by the bell code is to ask B if he is able to accept the train. If B is sure that any previous train has cleared the block section and that the line to a distance of ¼ mile in advance of his home signal (known as the clearing point) is also clear, then he may respond in the affirmative by repeating back the bell code. If B is not able to accept the train, perhaps because the line is blocked by another train, he will simply not reply. A must then try again a few minutes later. The extra ¼ mile to the clearing point is a safety margin in case the engine driver misjudges his braking distance.

Once B has indicated that the train can be accepted, he moves the pointer on his telegraph instrument to 'line clear' (see figure 2.6). This indication is transmitted back to A, whose telegraph instrument will repeat 'line clear'. Once A sees this, he will set any points that may be necessary, then pull the lever which clears his up starter signal. The driver, seeing this signal, will then get under way. As the train passes beyond the signal, the signalman at A will check that it is complete (by observing the red lamps on its rear vehicle), then send two rings on his bell to B to indicate 'train entering section'. On receiving this code, B will set his telegraph instrument to 'train on line', which indication is repeated on A's instrument so that A is aware that he cannot, for now, send further trains to B.

As the train arrives at B, the signalman there will pull off his 'up' home signal. As the train passes it, he will again check the train is complete, and if so send back to A a bell code to indicate 'train out of section', and at the same time set his

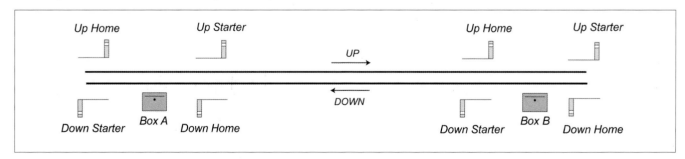

Figure 2.7 – *Block Section Working*

telegraph instrument back from 'train on line' to its central position, referred to variously as 'normal' or 'line blocked'. If the train is travelling onwards from B, then the signalman at B will have gone through the same procedure as did A, with the next signal box along the line, C say, so that by the time the train arrives at B it can be signalled on to C, and so on.

Clearly, this is a very much safer procedure than the time interval method of signalling. Provided that the signalmen at A, B, C, etc follow the rules, then even in the event of a train breaking down, collisions should not occur.

Single Line Working

In the case of minor routes, these were often single track, with passing places at stations or in isolated areas at passing loops. Between these passing places, the single track would serve as both 'up' and 'down' lines. It was therefore absolutely essential to ensure that trains did not enter the single line from both ends, since this would result in a head-on collision, with very much greater risk of death and injury than on a double track.

In the early days, a pilotman was employed between each pair of stations. His job was to travel between the stations on alternate trains. Drivers were not permitted to set off without the pilotman, who naturally could only be in one place at any one time, and therefore trains could not enter the single line section from both ends.

This, however, was very expensive in manpower and it was soon realized that the pilotman could be replaced by a wooden staff or token bearing the names of the two stations between which it was valid. Drivers could only travel on that section of line if they were in possession of the appropriate token, and since there was only one token, this again ensured that two trains could not enter the section simultaneously.

Although an improvement on the pilotman scheme, this procedure too was inefficient. There being only a single token, trains had to pass alternately in each direction. Two consecutive trains could not run in the same direction. Various schemes were devised to overcome this. In one such, the driver would be shown (but not given) the token and a ticket written out to indicate that he had seen it. The token would follow on the last train to run in that direction, whereupon it would be available at the other end of the single line. Even so, if a train was delayed, a train in the other direction could not take its place, because the token would be at the wrong end of the line.

In 1878, Edward Tyer (whose name we have mentioned already) patented his electric token machine which, along with several rivals' equipment doing much the same job,

Figure 2.8 – *Tyer's Token Machine*

came to define single line working (figure 2.8). Essentially, the machine was an electric telegraph instrument with an additional facility which allowed the signalman to draw out a token from the machine upon receipt of the 'line clear' indication from the box in advance. Once the token was withdrawn, no further tokens, from either this machine, or the one at the other end of the single line section, could be withdrawn until this token had been replaced in one or other machine. However, once the token was replaced, the next one could be withdrawn from either machine, thus allowing complete flexibility as to the direction of successive trains.

Lock and Block

At the end of our brief description of block section working, there occurs an enormously important phrase : *'provided that the signalmen follow the rules'* collisions should not occur. However, human nature being what it is, signalmen, through either forgetfulness, distraction, or whatever, will occasionally get things wrong. The key to safe block working is that the signalman at the entry to a block section should not clear his starter signal unless the line clear indication is showing on his telegraph instrument, and

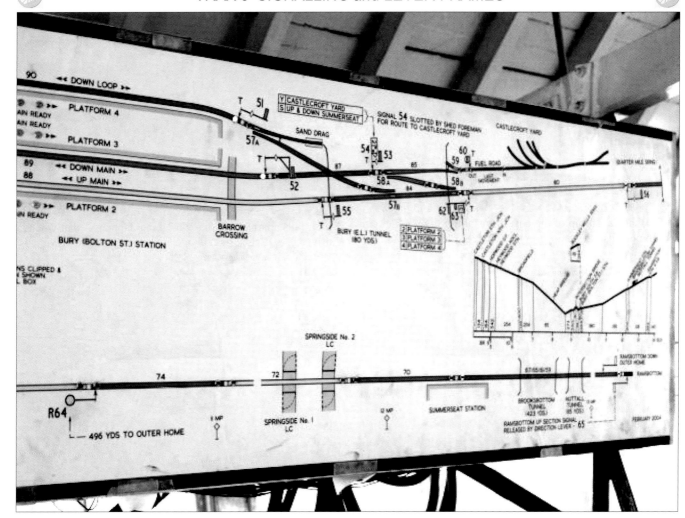

Figure 2.9 – *Track circuit indicator lights on the signal box diagram (next to signals 51 and 52)*

having done so, he should not signal a second train until the first is 'out of section'.

In 1874, William Robert Sykes invented an instrument that went some way to removing the human element from this procedure. First employed on the London, Chatham and Dover Railway, the Sykes 'lock and block' system again involved a modified telegraph instrument. To illustrate how the scheme worked, let us again suppose that we have two signal boxes, A and B, and that A wishes to send a train to B. As with standard block section working, A would call B's attention, then ask by means of an appropriate bell code if the line is clear.

B would reply to A as before, but with his Sykes instrument he would also press a plunger, thereby electrically unlocking A's starter signal. Until B had done this, A could not physically move his starter signal lever. Once the lever had been operated, clearing the starter signal at A, the train would set off, A would send 'train entering section' to B, and restore the starter signal to danger behind the train. This

starter signal would then again be locked at danger until the train had operated a treadle inside B's home signal, and the latter had been restored to 'danger'. Only then could A's starter be unlocked by B sending a second line clear indication.

Track Circuits

The Sykes scheme relied on mechanical treadles, operated by the wheel flanges as trains passed over them. A more sophisticated system was devised in America by William Robinson in the 1870s and first employed on the Philadelphia and Erie Railroad. This used a low-voltage battery wired across the rails of one end of an insulated section of line. Across the other end of the insulated section a relay was wired between the rails. Current from the battery passed down one rail, through the relay coil, and back along the other rail. This 'track circuit' current held the relay closed, and this gave a 'track clear' indication in the signal box. When a train passed over the insulated section, its wheels would short out the battery, thus removing the current from the relay which would then drop open, giving a 'track occupied' indication in the signal box. Originally, this would generally be a galvanometer needle, but later it would invariably take the form of a lamp which would light up on the signal box diagram (figure 2.9).

Sykes himself was interested in the system, and experimented with it during the 1870s and 1880s. The railway companies were reluctant to adopt it, however, partly at least because of the popularity of carriage wheels with wooden centres (Mansell wheels) which, being insulated by the wood from their axles, would not operate track circuit equipment. As ever, the cost of modifying all these wheels, on top of the cost of installing the track circuit equipment itself, was a prime consideration for the railway companies. There was a rule in the rule book, number 55, which required the driver of an engine held more than a few minutes at a signal to go, or send someone else, to the signal box to remind the signalman of his presence. The companies felt that this, if correctly observed, would achieve much the same result as a track circuit.

However, a number of fatal accidents served to demonstrate the need for a more reliable system than this. On Christmas Eve, 1910, at Hawes Junction on the Midland Railway's Settle to Carlisle line, two light engines had been waiting at the box's starter signal for the starter to drop. However, the signalman had forgotten they were there, and signalled an oncoming express as a through train, pulling off both his home and starter signal, causing the express to plough into the back of the two light engines with the loss of 9 lives. An even worse accident, in fact to this day Britain's worst, occurred in 1915 at Quintishill near Gretna and cost 227 lives. An up troop train collided with a stationary local passenger train, and then within one minute an express passenger train ran into the wreckage. Much of the coaching stock was wooden and gas-lit and a fierce fire took hold and burned for two days. The accident report laid the blame on two signalmen, one of whom had shunted the local train 'wrong line' to clear the down line for the express, then handed over the box to the man on the next shift, who signalled the up troop train into it, 'line clear' having been wrongly given to the box in rear. The report made the observation that, had the line been track circuited and the home signal thereby locked, then the collision might well have been prevented.

In the same year as the Quintishill accident, Arthur Bound, signal superintendent of the Great Central Railway, presented a far-sighted paper to the Institute of Railway Signal Engineers (IRSE), in which he called for the installation of track circuits, and the replacement of manual block section working by a system based on these. Where track circuits were employed, rule 55 was relaxed, and to indicate to the train driver that this was the case, a white diamond was attached to the signal post, as seen in figure

3.4. Bound's paper also called for the replacement of semaphore signals by colour lights, automatic signalling, and a means of preventing engines from passing signals at danger. We shall see how this vision of the future turned out to be remarkably prescient.

Cab Signalling

All of the safety devices we have discussed so far count for nothing without one important element. We can interlock points and signals, we can ensure that signals are only cleared when the section in advance is empty, and we can take any number of additional safety precautions. However, it all falls apart unless one vital thing happens – that when a driver sees a signal at 'danger' he stops the train. Signals passed at danger (SPADs in modern jargon) are not that uncommon. The driver may be distracted, the weather might make visibility poor, or the signal may be badly sited. For whatever reason it happens, a signal passed whilst at danger is an accident waiting to happen.

In 1894, Vincent Raven of the North Eastern Railway developed a mechanism for alerting an engine driver if he passed a signal set at danger. Just beyond the signal, and between the rails, lay a rod with two arms attached. When the signal was at clear, the two arms lay horizontal. When the signal was at danger the rod was turned by the signal operating wire so that the two arms were raised vertically. If an engine passed the signal, these arms tripped a steam valve which let steam at boiler pressure into a whistle inside the cab. The resulting ear-piercing blast would leave the driver in no doubt he had done something wrong! Furthermore, for trains fitted with suitable braking equipment, the brakes could be applied automatically by the trip mechanism.

Raven's apparatus was fitted to a number of lines on the North Eastern Railway; however, following the grouping, the LNER did not pursue it with much enthusiasm. It was the Great Western Railway that made much of the running in this field, following a bad accident at Slough station in 1900. The driver of a Paddington to Falmouth express went through a stop signal at danger and ploughed into the back of a passenger train standing in the platform at Slough station. The resulting collision caused 5 fatalities and 128 injuries, and prompted the development of what the GWR called automatic train control (ATC) which it went on to install on all of its main lines by the end of the 1930s. There is no doubt that this contributed significantly to the commendable safety record of that railway.

Automatic Signalling

The Liverpool Overhead Railway opened in 1893, and in many ways it was a wonder of its time. It ran for seven and a half miles, built on pillars above the roadway and not only

***Figure 2.10** – Liverpool Overhead Railway*

provided quick and easy access for passengers to Liverpool's docks, but also some spectacular views of the docks, liners, and the river Mersey. It was a considerable tourist attraction, and I can remember being taken as a small boy on family outings to the overhead railway before it closed in 1956. Its trains were all-electric, with power cars at either end and a trailer car in the middle. It was the world's first electric overhead railway, and, of more immediate interest in our current context, it was the first to be signalled automatically.

The power cars had trip arms projecting out over the trackside, and these made contact with lineside boxes, in which switch gear would operate the electrically powered semaphore signals. As the rear car of each train passed a signal, relays would set this signal to danger, and also reset signals further back to clear, thus achieving the same effect as block section working, but without requiring human intervention. In the early 20th century, the system was further enhanced by the addition of circuit breakers on the power cars, which would trip if the train passed a signal at danger and cut off the traction current. The electrically operated semaphores were also replaced around this time by colour-light signals.

The first main line to be equipped with automatic signalling was a stretch of the London and South Western Railway's Waterloo to Exeter route. In 1901, a new branch line was built from the small Hampshire village of Grateley to the military camp at Bulford, Wiltshire. Changes were required at Grateley station, and the opportunity was taken to re-signal the area using the (then) new concept of pneumatic power signalling. Instead of signals being operated mechanically by pull-wires, compressed air was used to move the semaphore arms. The most innovative aspect of the installation, however, was the fact that, on the main line between Grateley and Andover, track circuit relays were

used to operate the signals automatically, according the passage of trains. The installation was a success, and the LSWR went on to install similar schemes at Salisbury and, most notably, on the 24-mile stretch of its quadruple track main line between Woking and Basingstoke. The latter was completed in 1904 and gave service for over 60 years until colour light signals replaced it in 1967.

Route Setting

During the 1920s the GWR experimented with a new approach to signalling using equipment from the Siemens company. This was called route setting, and was first installed at Winchester Chesil station on the Didcot, Newbury and Southampton Railway (DNSR). In this scheme, a single lever was associated with each required route through the station and into the nearby goods yard at Bar End, which also accommodated a small engine shed and turntable.

The lever frame had just sixteen miniature levers. Each lever had four positions. In the normal 'off' position the signal associated with the lever was at danger. As the lever was pulled, it would be checked at an intermediate position in the frame. Relays would then sense the appropriate track circuits to ensure that they were clear, and, if so, electric current would drive the appropriate point motors to set whatever points were necessary. Once the points had been moved to the correct position and locked, the lever would be released and the signalman could pull it fully into the reverse position. At this point, current would be fed to the appropriate signal motor to move the signal to the clear position, signalling to the driver that he could now set off. After the train movement was complete, the signalman would push the lever back and again it would be checked at an intermediate position. The signal motor would be powered to return the signal to danger, any necessary points would be reset, and track circuits would again be automatically checked to ensure that the train move was complete. Once this was done, the lever was released, and could be pushed fully home into the normal position.

Although initially this scheme aroused great interest, the GWR only made one further installation, at Newport, Gwent, and the Winchester installation was replaced by a mechanical lever frame in 1933. After this, the GWR seemed to recoil from new innovations, preferring its own traditional way of doing things. On the others of the 'big four' railways, however, the new technology was developed with enthusiasm, and nowhere more so than on the Southern.

Southern Modernisation

At its creation with the grouping of 1923, the Southern Railway inherited an extraordinarily complex network of

Figure 2.11 – Winchester Chesil route setting lever frame

lines and stations in the capital, with a density of traffic that meant Victorian methods of working were barely able to cope. As well as the major termini of its constituent companies at Waterloo, Charing Cross, Victoria, Blackfriars, Cannon Street and London Bridge, the company had some of the busiest junctions in the world, at Borough Market for example, where the Charing Cross and Cannon Street lines merge on their way to London Bridge, and at Clapham, where the lines from Victoria and Waterloo meet on their way to the south and west.

At around the time of the grouping, the IRSE had set up a committee, chaired by none other than Arthur Bound, to examine the future of railway signalling and the options available. The USA had adopted three-position semaphores, but working in the upper quadrant so the arm was visible at all times (unlike the earlier 'slotted post' variety), and some railways had experimented with these. Others had looked at various different types of colour light signal. The committee's recommendations were published in 1924 and they were for the adoption of three- and four-aspect colour light signals. The four aspects were to be red - danger,

yellow - caution, double yellow - preliminary caution, and green - clear, these aspects being shown automatically as trains passed and triggering track circuits in advance of the signal.

The SR wasted no time in proceeding to adopt these latest innovations, converting their motive power to electric traction, and replacing their old signalling systems with the most modern available. Old mechanical lever frames, semaphore arms and Sykes 'lock and block' systems gave way to power frames, colour light signals, and track circuit block working. A 'power frame' is a lever frame that does not operate points and signals by direct mechanical linkages. Instead, the levers switch electrical power on and off and a remote electric motor or pneumatic device operates the point or signal. Power frames could therefore operate points and signals at a far greater distance than would be possible with mechanical linkages. Many smaller boxes could thus be closed down and their work transferred to larger centres.

Westinghouse style 'K' power frames were installed at Charing Cross and Cannon Street in 1926, at Borough

Figure 2.12 – Westinghouse type 'L' Power Frame at Clapham Junction 'A' Box

Figure 2.13 – 'NX' panel from Brunswick Goods

Market Junction in 1928, and a monster frame at London Bridge the following year with 311 levers. Behind each signal lever, four indicator lights (red, yellow, green, second yellow) would show what aspect the colour-light signal was showing. A signal lever could be pulled (locking permitting, of course) but it would only set the signal to clear if subsequent track circuits proved the line was unoccupied. The signal would be set back to danger automatically after a train had passed it. During the following decade, SR installed Westinghouse style 'L' frames, which were outwardly similar to the style 'K' frames, but incorporated electrical locking using relays instead of mechanical locking. These installations included Waterloo and Clapham Junction in 1935 and Victoria station and Battersea Park Junction in 1937 (see figure 2.12).

The End of Lever Frames

The early power frames built by Westinghouse and others were designed for a breed of signalmen who had, since the middle of the nineteenth century, been accustomed to mechanical lever frames. These had to be heavy duty items. Point blades, sometimes a couple of hundred yards away, together with all that point rodding, had to be moved by the signalman's muscle power alone. Therefore, plenty of leverage was required, and the levers were several feet long with handles that allowed for a two-handed grip. They were arranged in a row, with quadrant plates at floor level enabling the signalman's feet to get a good purchase. The miniature levers in Westinghouse and other power frames were designed to look exactly like their full-size predecessors, only smaller since they didn't need the leverage. They were still given little catch handles, painted

red, black and so on, placed in a row, and numbered left to right. Given that all they had to do was open and close electrical contacts, this was entirely unnecessary. The same could be done with a switch that could be held between finger and thumb.

At Thirsk, on the LNER, a resignalling project was undertaken in 1937, under signalling supervisor Arthur Ewart Tattersall, which broke the mould. Instead of a row of miniature levers, Tattersall had thumbswitches mounted on a panel upon which was drawn a track diagram. The switches were mounted adjacent to the signals to which they referred. Where a signal might apply to several different routes, there would be a bank of switches, one for each route. Throwing a switch would set in motion a sequence of events. Electric relays would check that the route selected by the switch was not in conflict with any route already set, and that all of the track circuits along the route indicated 'clear'. If so, points were set as needed, locked, and finally the appropriate signal would be cleared. This was the first panel employing what would come to be known as the 'One Control Switch' (OCS) system.

Although many OCS panels were built, there was a rival approach, whose first appearance was in 1939 at a very minor station called Brunswick Goods in Liverpool. This was a Cheshire Lines station, at this time a joint LNER/ LMS operation. At Brunswick was installed the world's first 'Entrance-Exit' (NX) panel, built by Metropolitan Vickers. This, like the OCS panels, had a track diagram upon it. At each signal location was a push-button, and all a signalman had to do was push the button adjacent to the point a train was to enter, then push a button adjacent to the point the train was going to exit (or halt), and relays did the rest. It was similar to the OCS panel, except that the signalman didn't even have to decide which route to set.

Over the years, a great many NX panels have been installed, however these too are now being overtaken by Integrated Electronic Control Centres (IECCs), where computer software has taken over not only the functions of the lever frames of dozens of separate signal boxes, but the functions of the signalmen themselves. The Automatic Route Setting facility allows the entire timetable to be fed into the computer system, which then drives the railway accordingly. Signallers intervene only when something out of the ordinary occurs. However, fascinating though this subject is, it is beyond the scope of this book. We have to draw the line at some point, otherwise the book would never be finished. We shall therefore confine ourselves to signalling by means of manual levers or switches.

3

SIGNALLING PRACTICE

We now take a rather more detailed look at the different types of signal and their purpose, and how they were used in some typical prototype track layouts. Note that signalling practice varied somewhat from one railway to another. This variation is more marked in the early days. Gradually, either through a process of evolution similar to Darwin's 'survival of the fittest', or because of an edict of the Board of Trade, railways began to conform to common standards. By the time of British Railways, signalling practice was reasonably standard throughout the country (except, in some respects at least, on the Western region). If you are modelling an early railway, and want to conform with the prototype exactly, you should consult some of the specialist books or web sites listed in Appendix 1.

Semaphore Signals

There was considerable variation in the painting of early semaphore arms. Those on the Midland Railway were red with a white spot; on the London & Brighton they were white with a red stripe. However the scheme used by most companies, and which became the standard, was to paint them red with a white stripe. So that drivers would not mistake a signal facing in the opposite direction, the back of the arm was painted white with a black stripe. Some examples will be found in the colour plates section.

As has been mentioned, the first semaphore signals had three aspects: 'danger', 'caution' and 'clear'. These indicated to the driver whether he was to stop, proceed with caution, or proceed at normal speed. However, the use of three-aspect signals was gradually dropped, and companies increasingly employed signals with just two aspects: 'danger' and 'clear'. The latter was indicated by the signal arm in the 45° position, rather than slotted inside the post. This had the not inconsiderable advantage that an arm at 'clear' could be distinguished from one that had fallen off its post! Sometimes, a signal at 'danger' might be referred to as being 'on', or when at 'clear' being 'off'.

Where there was a need to give the driver a warning, a 'distant' signal was used. This indicated the state of the 'stop' signal ahead. If the stop signal was at 'danger', then the distant signal too was put to 'danger'. This indicated to the driver that he should proceed at such a speed that he would be able to stop at the next signal. Initially, distant signals were painted exactly the same as stop signals – distinguishing one from the other was down to the driver's knowledge of the line. In 1872, the London, Brighton &

South Coast Railway began to cut a notch out of the end of their 'distant' arms, so that they could be distinguished from 'stop' arms. Both were painted red with a white vertical stripe. During the early 1870s, other railway companies followed suit, and by 1877 the Board of Trade insisted upon it. The lamp colours for stop and distant signals were identical: red when the signal was at 'danger' and green

Figure 3.1 - *Stop and Calling-on signals. (LMS)*

when it was 'clear'. It was not until the early part of the twentieth century that the Great Central Railway began to use a yellow light instead of red in their distant signals when in the 'danger' position. They also began painting their distant signal arms yellow with a black chevron matching the notch in the end of the arm. This made stop and distant signals much easier to distinguish, and other companies quickly followed suit.

In the 1920s, as we saw in the previous chapter, an IRSE report recommended the adoption of colour light signalling for main lines, and rejected the use of three-position upper quadrant signals. This left the way open for two-position semaphores to use the upper quadrant without danger of confusion. So after this date, upper quadrant arms began to replace those that operated in the lower quadrant. That is, instead of 'clear' being indicated by a signal arm inclining downwards, the arm inclined upwards at 45° or so. In both cases, 'danger' was indicated by the arm pointing horizontally. Upper quadrant arms had the advantage that the arm would naturally fall down into the 'danger' position should a mishap such as a broken wire occur. The lower quadrant arms had to be provided with heavy counterweights to achieve the same objective. Upper quadrant arms eventually replaced lower quadrant ones virtually everywhere except on the Great Western.

At junctions, where the running line diverged in two directions, separate signals were employed to control movements in each direction. They were often mounted on a bracketed post. When the diverging line was a branch off the main line, where perhaps a lower speed limit applied, the signal for the main line was mounted on a higher post than that for the branch. Where the diverging lines were both main lines with similar speed limits, the semaphore arms were mounted at the same height.

Running Signals

Stop and distant signals on main lines are referred to collectively as running signals. Their function is to facilitate the safe running of trains up and down the lines. We saw in the previous chapter how a typical signal box might have two stop signals for each direction, a home signal in rear of the box, and a starter signal in advance of it. However this arrangement was not universal. Where traffic was light, the two signals might be rolled into one, as it were, and the box would have a single stop signal for each direction. This was often the case where a signal box was not associated with a station, but was in an isolated spot, its purpose being simply to break up an otherwise over-lengthy block section. Trains accepted from the box in rear would have to come to a halt at the signal, unless they had already been offered to and accepted by the box in advance, in which case the signal could be cleared.

Figure 3.2 - A (fixed) distant signal. (SR)

On the other hand, if the line were a very busy one, there might be more than two stop signals. These would be referred to as (in order) the outer home, the inner home, the starter and the advanced starter signals. The purpose of having several stop signals under the control of one box would be to allow for greater throughput of traffic. If the box were at a station (as would almost invariably be the case) and supposing it had a single home signal within ¼ mile of the platform (i.e. the platform was within the clearance point), then a train could not be accepted from the box in rear until any train in the platform had allowed its passengers to leave and board and had got under way, having been accepted by the box in advance.

By having an outer home as well as an inner home, the signalman could accept a train from the box in rear, lowering his outer home signal, but keeping the inner home at danger until the previous train had departed. Likewise, having two starter signals would allow him to lower the

Figure 3.3 - *LBSCR Stop and Shunt-ahead signals at Sheffield Park.*

the distant signal referring to box B's home signal on the same post, and beneath, box A's starter signal. The situation could then arise that A's starter was set to danger, but B's signals were clear. If the distant signal on the post with A's starter were cleared this would present to the driver a somewhat confusing picture of a stop signal at danger with a distant signal at clear.

To prevent any such confusion, the distant signal would be slotted. That is to say, the 'distant' lever would have to be pulled off on B's lever frame, and also A's starter signal would have to be pulled off, before the signal would clear. A slotting mechanism at the base of the signal post would ensure this.

On several railways in the early days, stop signals referring exclusively to goods lines employed a slightly shorter semaphore arm which, instead of a white stripe, had affixed to it a white circle. The signal was, apart from its appearance, exactly the same as a standard stop signal: if it was at danger it meant 'stop'. The purpose was solely to aid drivers in distinguishing which signal applied to their train.

Subsidiary Signals

To control shunting and other movements within stations, a variety of subsidiary signals were used. Again we emphasise that practice varied from railway to railway, and also over time.

The Shunt-Ahead Signal

In the previous chapter, we saw how various schemes were employed to ensure that only one train at a time could be in a block section. However, at certain stations, depending on layout, it may be necessary for a goods train to draw forward beyond the station's starter signal before backing into the goods yard. The rules dictate that the main starter signal could only be cleared for this move if the complete block section up to and including the clearance point was verified as clear by the signal box in advance, and 'line clear' was given to release the starter signal. However, this would be unnecessarily restrictive if all that was required was to draw the train forward a couple of hundred yards.

In such circumstances, a shunt-ahead signal could be provided. This would generally be on the same post as the main starter signal, and would take the form of a slightly shorter arm with a metal casting of the letter 'S' attached to its face. Figure 3.3 illustrates a London, Brighton and South Coast example. Some distance beyond the signal, a notice board would be fixed with the legend 'Limit of Shunt'. If the shunt-ahead signal were set to clear, engines would be authorised to proceed past the starter signal, even though this was at danger, but only so far as the limit of shunt, prior to reversing into the goods sidings.

platform starter, thus clearing the platform, but holding the train at the advanced starter until the signalman at the box in advance was ready to accept it. The most advanced stop signal is often referred to as a section signal, since it is this signal that controls entry into the next block section.

In situations where a signal box controlled more than one stop signal, the rule was that the distant signal preceding the first stop signal could only be cleared if all the stop signals controlled by that box were clear. Locking on the lever frame could ensure this. Thus, the driver would know that, if the distant signal were clear, he had a clear run through and could set his speed accordingly

In certain circumstances, where boxes A and B were particularly close together, it might be necessary to mount

Figure 3.4 - *LMS Stop and Calling-on arms with an alphanumeric direction indicator.*

Before clearing the shunt-ahead signal, the signalman was required to warn the box in advance by sending a bell code (3-3-2) signifying 'shunting into forward section'. The signalman at the box in advance would then set his block instrument to 'train on line' until such time as the shunt was complete, indicated by receipt of a bell code (8 consecutive beats) from the box in rear.

The Calling-On Signal

Generally, when a signal was changed from the danger aspect to clear, it meant that the line ahead was clear and the engine driver could proceed. The calling on signal would be used in circumstances where an engine was required to pull forward even though the track ahead was occupied. There might be a number of reasons why this would be necessary.

For example, certain station platforms were long enough to accommodate more than one train. In this case, we might

already have one train in the platform and wish to signal another one in behind it. Alternatively, at a terminus station, a train might have arrived and after the train engine has been uncoupled, we wish to signal a station pilot engine to enter the platform to draw the coaches out of the platform and place them in a carriage siding. In these and similar situations, a calling-on signal would be employed.

This is typically a smaller version of the stop signal, and in the early days would be painted red with white letters 'CO' painted on it. Later this was replaced by painting a white horizontal band on the signal, as in figure 3.4. This style of painting was gradually adopted for both the shunt-ahead 'S' and the warning signal (see below). When the subsidiary arm moved to the clear position it would reveal a panel containing the letter 'S', 'C' or 'W' to indicate exactly what type of subsidiary signal this was.

The Warning Signal

We have seen how the rules for block section working were designed to ensure that only one train could occupy a block section at any one time, and that before a train was allowed to enter the section, both the section itself and a quarter of a mile clearance in advance of it must be empty. However, in the case of freight trains a little more flexibility was allowed in certain circumstances. For example, a goods train might enter a block section which had an industrial siding part way along it, where wagons might need to be dropped off or picked up. If both the block section and the ¼ mile clearance beyond the home signal in advance had to be closed to any other traffic whilst these shunting operations took place, then considerable delays could result.

Therefore, regulation 5 of the rule book permitted a train to be accepted into a section under what was called the 'warning arrangement'. The section signal would be kept at danger until the engine was almost at a standstill, then it would be lowered and a green flag (or light at night) would be shown to the driver by the signalman. The driver would confirm that he had seen this by a short blast on his whistle. This indicated that he understood that his train had a 'warning acceptance', and this meant he had to proceed with caution because the line might not be clear to the clearing point.

Where this type of working might be fairly frequent then a special subsidiary signal, called the warning signal, would be employed. This was generally mounted on the same post but below the main section signal. When cleared, it permitted a train to enter the section even though the main section signal was at danger, and the driver would understand that he was proceeding under the warning arrangement.

A similar circumstance applied where 'permissive block'

Figure 3.5 - *A Shunting signal (SR)*

working was allowed. The 'absolute block' system whereby only one train was allowed in a section applied generally to passenger lines. However, on the approach to large marshalling yards, for example, if a queue of goods trains developed waiting to enter the yard, and each occupied a whole block section, this would cause considerable delays. In such circumstances permissive block working, whereby several trains could occupy a single section whilst they waited, was employed. If the section happened to be empty a train would be signalled into the section using the normal section signal. If not, it would be accepted under permissive block conditions and the main section signal kept at danger, whilst the subsidiary warning signal would be cleared.

The Backing Signal

This appears to have been a signal pretty much exclusive to the GWR. The signal arm was plain red with no white stripe, and had two small holes in its face. The rear of the arm was white with a small black stripe. Its purpose was to signal a move in the 'wrong direction' on track within station limits. In the early days, the LSWR had a 'shunt wrong road' signal which served a similar purpose. This was shaped like a bow-tie with two longitudinal struts

crossing in the middle. However, unlike its GWR counterpart, it did not last. Shunting signals in the form of ground discs generally took over this purpose.

Shunting Signals

Shunting signals control the movement of engines in and out of sidings, or through crossover points during shunting operations. Therefore, in general, the engine will be travelling very slowly and does not require a long braking distance. The signal need not, therefore, be visible from afar and consequently need not be up on top of a tall post. Most shunting signals were at ground level (though see figure 3.6) and took the form either of a miniature version of the semaphore stop signal, or of a white disc with a red stripe. The disc eventually superseded the miniature semaphore on virtually all railways. Like the semaphore signal, the red stripe is horizontal for 'Danger' or inclined diagonally for 'Clear'. Where such signals controlled a number of routes, for example at the entry to a 'fan' of sidings, they were often mounted one above the other on a short post. In such cases, the topmost signal always referred to the leftmost route, the next one down to the route second from left, and so on.

Figure 3.6 - *Not all discs were on the ground. This SR signal is mounted on a post for visibility*

In some situations, the shunting disc signal might have a yellow stripe instead of a red one. This is not a 'shunting distant' signal, however! The 'yellow disc', as it was called, was a special shunting signal for use in goods yards. Suppose we have a typical goods yard with a fan of sidings and a headshunt. Usually, individual shunting moves between the sidings would not be controlled from the signal box. Instead, points giving access to the sidings would be operated from a separate lever frame within the yard, known as a ground frame. Not infrequently, the ground frame would control points only, and shouted instructions would replace formal signalling. However, there would be at least one signal and set of points giving access from the goods yard to the running line(s), and this would need to be under the control of the main signal box, so that it could be suitably interlocked with other points and signals. During shunting moves, this signal might have to be passed many times, as the engine went back and forth between sidings and headshunt, and for the signalman to have to clear it each time would be tedious to say the least.

In such situations, a yellow disc would be employed. The rule for yellow discs was that they could be passed when set at danger provided that this was in a direction other than that to which they referred. Thus shuffling back and forth past them between headshunt and sidings was permitted, but the signal would have to be obeyed if the movement were out of the yard and onto the running line. Because of the difficulty, under poor lighting conditions, of distinguishing a yellow stripe on a white disc, the discs were painted black in later years.

Siting of Signals

The general rule was that signal posts were sited on the left

hand side of the track, at the location where the engine was required to stop. The semaphore arm extended to the left hand side of the post. However, in some situations the left hand side of the track was obstructed in some way, or it might have been obscured from view until the engine was right upon it. In such circumstances the post might be moved to the right hand side of the track, or even to the right of the track beyond.

To aid visibility from a distance, signals might be mounted on tall posts. In locations such as just beyond an overbridge, it may have been that an extra tall post was required for visibility from a distance. However, once the engine had reached such a signal, the driver would have difficulty seeing it from inside his cab. Therefore, a second semaphore arm might be added lower down the post for visibility from close to. This was known as a co-acting signal, and would be operated by the same wire as the main signal.

Sometimes, it was impossible to give reasonable visibility from a distance to a signal. This would be the case if the signal were on a sharp curve in an embankment, for example. In this case, a repeater could be employed. This most often took the form of a white disc with a black band. Unlike a shunting disc, it would be mounted on a post. With the band horizontal, it indicated that the signal ahead was at danger. With the band diagonal, the signal ahead was clear.

Where visibility was made difficult, not by intervening structures etc. in the foreground, but by a cluttered background, a white panel would be fixed to the post behind the signal. On station platforms it was sometimes necessary to position signals under awnings. If there was not much headroom, such a signal might have a shorter arm, pivoted at its centre, rather than at its right hand side, so as to minimize its vertical swing.

Colour Light Signals

We have seen in the previous chapter how Bound's committee recommended the use of three- and four-aspect colour light signalling. The LMS, LNER and SR adopted these recommendations and we have seen how the SR was quick to implement them for its suburban routes into south London. The LNER soon followed suit as, later, did the LMS. There were detail differences, but basically all three companies adopted a design similar to a road traffic light. The coloured lenses were mounted one above another on a vertical rectangular box, with hoods to improve visibility. Unlike road traffic lights however, the colours read, from the bottom, red, yellow, green, second yellow (if required). These meant: red – 'danger, stop', yellow – 'caution', double yellow – 'preliminary caution', and green – 'clear'.

The GWR, unsurprisingly, decided to be different. They were far from enthusiastic in their installation of colour lights to replace their (lower quadrant) semaphores. In those cases where they did so replace, the colour lights were made to look exactly the same as semaphore signals would look at night. Instead of a simple red, yellow, green traffic light, the GWR colour light signals mimicked a semaphore stop signal with a distant mounted on the same post below it. The upper light would be able to display the colours of a stop signal spectacle plate, i.e. red and green. The lower light would display those of a distant signal, i.e. yellow and green. Thus 'stop' was indicated by red over yellow, 'caution' by green over yellow, and 'clear' by green over green. Multiple aspect lights were not used. Instead, a white light was used, with a semaphore-like spectacle swinging in front of it to give the red, green and yellow colours.

The earliest colour light signals, like those on the Liverpool Overhead Railway, displayed just two aspects, red for 'danger, stop' and green for 'clear'. Where operated by a signalbox (they were automatic, of course, on the LOR), the lever was pulled to clear the signal and restored to set it to danger, exactly like a semaphore.

However, with three- and four-aspect signals, the situation is a little more complicated. A simple lever pull could not cover 'danger', 'caution', 'preliminary caution' and 'clear' indications. Such signals are, therefore, generally operated by track circuits. When a train passes a green light, the green will change to red. As it passes the next track circuit, the red light will change to yellow, as the train passes the next one along, yellow will change to either green or double yellow, depending on whether the signal is three- or four-aspect. If, in the meantime, another train passes the same signal it will, of course, revert to red.

Direction Indicators

With colour light signals at junctions, the SR and LNER in the early days followed the same practice as with semaphores, that is, two separate sets of colour lights were employed – one for the 'straight on' route and one for the diverging route. These were mounted on a bracketed post exactly like semaphores, with the higher speed route on the taller post.

However, this practice was superseded by the use of direction indicators, colloquially called 'feathers'. These were mounted above the colour lights and consisted of a line of white lights inclined at 45° upwards to the left or right to indicate that the route was set for the diverging line. The SR used a line of three white lights, the LNER five. Five lights eventually became the standard. If there was more than one diverging line, then the second was indicated by a horizontal line of white lights to the left or right, and the

Figure 3.7 - Three-aspect colour-light signal at Silecroft, Cumbria.

third by a line inclined downwards at 45°. Where the route was set for 'straight on' none of the feathers would be illuminated.

On lower speed lines, and at stations, alphanumeric direction indicators can be used. These comprise a large panel upon which an illuminated letter or number is displayed to indicate the route set, as in figure 3.4.

Subsidiary Colour Light Signals

The ground discs described above continued in use alongside colour light signals in many cases. They were illuminated at night, and in this situation, the yellow band on a white disc was difficult to see. Therefore, following a lead set by the SR, for yellow discs the white background colour was often replaced by black.

However, colour light shunting signals did appear in many

Figure 3.8 - Four-aspect colour-light signal with direction (feathers) indicator.

cases replacing discs. The early types were miniature two-aspect signals displaying either green or red. However, the favoured type was a triangular plate upon which were mounted three small lights arranged in a triangle. At the right was a white light. To its left, and in a horizontal line with it, was a red light, and above this was a second white light. To display 'danger' the red light and right hand white light were lit. To display 'clear' the white lights were lit. In a later version of this signal, the right hand light was a bi-colour one. It shone red when the signal was at danger, or white when clear. Thus the driver would see two horizontally aligned red lights for danger, or two white lights inclined upwards for clear. Like the shunting discs they replaced, these signals were generally placed at ground level.

Similar arrays of red and white lights could also be mounted

on the same post as a colour light running signal as a calling-on or shunt-ahead subsidiary signal. Their meaning was exactly the same as their semaphore counterparts, and again a small panel with 'S' or 'C' would be included to indicate whether this was a shunt-ahead or calling-on signal.

Some Typical Examples

Having looked at the different types of signal, we shall now look at some practical examples of signalling using typical track layouts. The examples are (or were) actual stations or junctions, and the signalling is as per the prototype. For various reasons that will be discussed in the next chapter, it is not always practicable to signal a model exactly like the prototype. In the next chapter, therefore, we shall look at ways in which the signalling can be adapted to fit within the constraints that most of us have to observe for our model railways.

Weyhill – A Through Station

The first example we take (figure 3.9) is that of Weyhill, a small double track through station on the Midland and South Western Junction Railway. The latter was conceived as a link from the Midland Railway's metals to those of the London and South Western Railway, thus giving access from Birmingham to Southampton. Weyhill was a station on the original Swindon, Marlborough and Andover Railway, which became part of the MSWJR.

We first note that, in accordance with standard practice, each signal and each set of points is allocated a number. The sequence of these numbers is not random – it conforms to a logical pattern which most layouts would follow. Imagine yourself to be in the signal box, then the levers controlling running signals on the line on which trains travel from your left to your right are placed on the extreme left of the frame, with the furthest left being lever number 1 (in Britain at least, trains, like cars, go on the left). Levers controlling running signals in the opposite direction are on the extreme right of the frame, and so are numbered (in this case) from 19 downwards. The logic behind this is that it groups the levers together so that the signalman can signal a train through the station without having to constantly walk to and fro along the lever frame. Lever number 1 is the down distant signal, 2 is the down home, 3 the down starter and 4 the down advanced starter. To signal a down through train, therefore, the signalman would pull 2, 3, 4 and 1 (as we shall see later, locking would prevent him pulling 1, the distant, until the other three were 'off'). Notice that signal 2 is on the 'wrong' side of the lines. As the station was built on a slight curve, this would aid visibility.

In the other direction, the lever on the extreme right, number 19, is the up distant, 18 the up home, 17 the up starter and 16 the up advanced starter. Notice that signals number 4

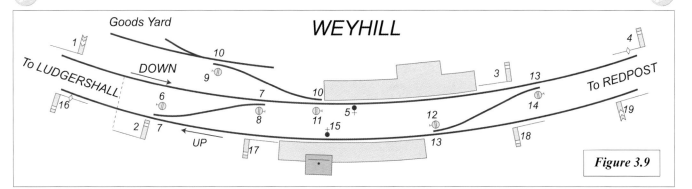

Figure 3.9

and 16 on the diagram both have a small diamond shape on their post. This indicates that the line adjacent to the signal is track circuited. A similar white diamond shape on the actual signal post would indicate to the engine driver that this was exempt from rule 55, that is, the fireman or guard does not have to walk back to the signal box to remind the signalman of their presence if the train is held there. The introduction of track circuits here must have been welcome, especially in the case of signal 4, since the round trip to the signal box and back would have been almost a mile!

Between the levers for the running signals at either end of the frame are the levers for the crossover points and their associated ground discs. Notice how these, too, are grouped together. Lever 7 works a pair of points forming a crossover at the north end of the station. Lever 6 works a ground disc giving access from the up line to the down over this crossing. Lever 8 works a disc for access from down to up. Thus, each of these moves requires the signalman to pull just two adjacent levers, 7 then 6, or 7 then 8. The points must be set before the signal lever can be moved – locking would ensure this. Levers 12, 13 and 14 operate a similar crossover and its associated discs at the south end of the station.

Between these crossovers, points 10 give access to the goods yard. Ground disc 11 signals a move from the down line into the yard, disc 9 signals a move back from the yard to the down line. Within the goods yard itself, any necessary points would be set manually by the shunter. The two remaining levers, 5 and 15, are detonator placers, used in foggy weather. When either lever is pulled, a detonator is

placed on the rail. If a driver fails to see one of the stop signals because of poor visibility, he will go over the detonator and the resulting loud bang will alert him to his error.

Red Post Junction – A Junction

A short distance from the previous station, the MSWJR formed a junction with the LSWR main line from Andover to Grateley (where, you may remember from the previous chapter, the first stretch of main line was automatically signalled). The track diagram (figure 3.10) shows the junction as it was in the early 1880s. The line from Weyhill was single by the time it got to here, and the layout is typical of how a single line formed a junction with a double.

Signals 1 and 2 are the junction distant and stop signals for the down main line. On the same posts, signals 3 and 4 signal a move from the down line over the junction and onto the branch. For this latter move, points 6 must first be set, and since they are in a facing direction, a facing point lock (FPL) is provided. This is worked by lever 5, the lock being represented by the short bar on the diagram.

In the up direction, there is no access to the junction. The main line is signalled by a distant, 12, and stop signal, 11. Notice that points 7, being trailing, do not require a FPL. From the branch onto the up line, points 7 must be reversed, but points 6 are left normal. However, these being facing, the FPL, lever 8, must be reversed before signals 9 and 10 can be pulled 'off'.

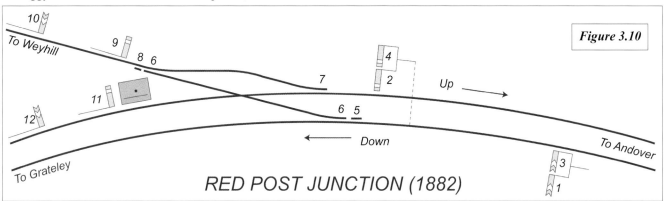

Figure 3.10

RED POST JUNCTION (1882)

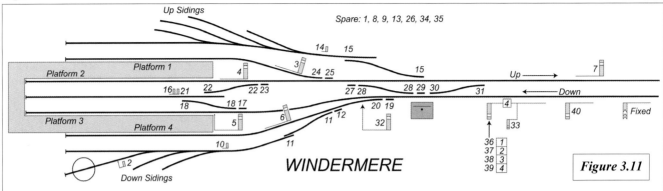

Figure 3.11

Windermere – A Terminus Station

Windermere is the terminus of a branch that leaves the West Coast Main Line at Oxenholme. It was authorised in 1845 as the Kendal & Windermere Railway, and opened in 1847. In its heyday, such was the number of train-loads of visitors that no less than four platforms were provided. Sadly, today the station has but a single platform. Figure 3.11 shows a (slightly simplified) layout of the station in happier times. Looking first at the incoming (down) line, we have a fixed distant then an outer home signal (40). The inner home signal is worked by four levers, 36-39. In addition to pulling off the signal arm, each lever would cause a different number to be illuminated in the direction indicator box, as in figure 3.4, to tell the driver what platform he was signalled into. Points 20, 24 and 28 would select the platform, and because these are all facing they have facing points locks worked by 19, 25 and 29 respectively. Once the route was set and locked, the appropriate one of levers 36-39 would be released.

In the outgoing (up) direction, each platform had its own starter signal, these are numbers 3, 4, 5 and 6. Again, points would require to be set and locked appropriately before any one of the starters could be released. Signal 7 is an advanced starter, which would allow the signalman to move an outgoing train from one of the platform roads even if the next signal box (Staveley) was not yet able to accept it, thus freeing up the platform for an incoming train. That such an arrangement was thought necessary is a testament to the popularity of rail excursions to the Lakes in years gone by!

On the down side there was a turntable and a series of carriage sidings. There was also a carriage siding in the middle of the station, between platform roads 2 and 3. Access to this was via points 18 and 22. On the up side, there were further sidings used for goods traffic. The sidings points themselves were not worked from the signal box, but from ground frames, although access to the running lines was controlled by points 11 and 15 and signals 10 and 14 respectively.

Whitchurch – A Single Line Station

Whitchurch (Hants) was a station on the Didcot, Newbury and Southampton line. South of Newbury, this was a single line worked by Tyer's key token. It is, in fact, the prototype for the author's layout. The diagram in figure 3.12 represents the station layout after some improvements were made in the 1940s.

The running signals follow the standard pattern. In the down direction, there is a distant signal fixed at danger as in figure 3.2. This is because even a through train would be required to slow down to no more than 15 m.p.h. for the token exchange equipment. Following this are the down home signal (1), starter (2) and advanced starter (3). In the up direction there is again a fixed distant, then up home (27) and starter (26).

The main points for the passing loop at the south end of the station are operated by lever 18. Notice how this is arranged as a pair of points, one of which switches the route through the station to the up or down line, the other of which serves to de-rail any vehicle that might be running towards the points from the station area whilst they are wrongly set. These are referred to as catch points and are there to try to

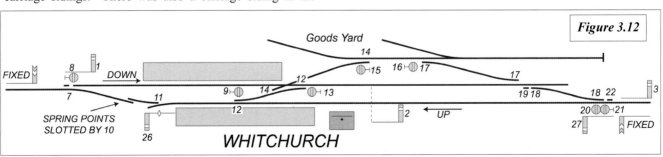

Figure 3.12

prevent runaway vehicles from getting onto the main line. Such an accident happened to a BR Standard Class 4 in 1953, when it was unable to stop a heavy goods train and was de-railed at points 18. Notice that these being facing points, they are locked by levers 19 and 22.

At the other end of the loop, points 7 and 11 provide a similar function. However, these are operated in a different manner from points 18. At around 600 yards from the signal box, they are too far for a manual pull, and are thus motor operated, electricity being supplied from a hand generator in the box. The motorized points incorporate their own locking function and so do not need separate FPL levers. Notice the spring points. If a train runs over them in the correct direction, the blades, being free to move, will simply be pushed over by the train wheels. However, if a vehicle runs over them in the 'wrong' direction (which could happen if a vehicle became uncoupled and ran downhill back towards the station), the points will ensure the vehicle is de-railed before it gets into the station area. If a shunting move is required 'wrong line' over the spring points, then lever 10 could be used to set them accordingly.

Levers 9, 12 and 13 control a crossover from the up to the down line and vice versa. Pulling lever 14 in conjunction with 12 gives access from the up line to the goods yard, levers 9 and 15 signal moves over this crossover. Notice that none of these points require locking bars, as they are all trailing for a train running in the normal direction. A similar crossover, 17, gives trailing access from the down line to the yard. During shunting operations, it may be necessary to run 'wrong road'. Therefore, ground signal 8 provides access to the 'up' line from the north end. Signals 20 and 21 provide access to the down line and yard respectively from the south end. Lever 16 signals a move from the yard onto the running line.

The frame in all has 27 levers, although only 21 are used. Positions 4, 5, 6 and 23, 24, 25 are spare. Points within the goods yard are worked by a small ground frame.

'Fictitious' Layouts

If your desire is to model a specific station, then you will probably be able to find an appropriate track plan and signalling diagram. Some of the web sites listed in Appendix 1 may be able to help. However, if your model is of a fictitious location then you will need to devise both the track plan and signalling yourself. In reviewing the typical signalling arrangements above, we have highlighted some general principles. Even a free-lance layout ought to conform to these, so that it has an authentic 'look and feel'.

Make sure that wherever possible points off running lines are trailing rather than facing. Where this is impossible, for example on single lines, the prototype would have facing point locks. Not many modellers actually bother with working FPLs, although dummy FPL castings can be bought and add to the layout's realism.

Running line signals should be sited in a manner consistent with block section working. Few of us have sufficient space to have scale length block sections, but your signals should obviously be further apart than your longest train and should follow the standard practice outlined above.

Subsidiary signals should be provided for all movements onto and off running lines, although within a goods yard or a fan of carriage sidings where points are worked by a ground frame it is not normally necessary to signal each and every shunting move.

With regard to numbering points and signals, standard practice was to assign the outermost levers of the frame to the furthest signals on either side of the signal box, then to work inwards from either end. Thus, a typical sequence might be (1) distant, (2) home, (3) starter, (4) advanced starter. Around the centre of the frame would be the levers assigned to points and signals within station limits. Notice how these are grouped in the above examples so that wherever possible points and signals for a particular move are close together on the frame.

Above - Real or model? It is of course a full size frame (mechanical locking in the tray on the floor) but operating a model! (Left) The layout was in the former Southern Railway signal school at Clapham Junction. Unintended realism was created as smoke from engines passing underneath would percolate up through various holes in the floor!

4

SIGNALLING A MODEL

There are a number of reasons why a model might be signalled somewhat differently than the prototype. Of course, there are some models which are exact scale representations of a station track plan, and they can be signalled exactly as per the real thing. Indeed, some model railways are signalled using actual preserved prototype lever frames – have a look at Mark Adlington's Westinghouse web site listed in Appendix 1 for example.

However, for various reasons such an approach is impractical for most of us, and even if it were possible it might not be the best way of going about things. The most common problem is one of space. My own 0 gauge model is of Whitchurch station, as described earlier. The two home signals (Nos 1 and 27) were approximately half a mile apart on the prototype, which in 0 gauge would be around sixty feet. Very few of us have this kind of space available, so in my case, this distance is telescoped to a mere sixteen feet. This can often be done without sacrificing too much of the operational possibilities of the layout simply by making sidings and loops shorter. Goods trains of thirty or forty wagons will have to be replaced by half a dozen wagons or so, but in principle at least, train movements will be similar.

Space is not, however, the only consideration. You should also give some thought to the operational interest of the layout. Many stations, particularly the smaller ones of a type which lend themselves to small-scale models, were actually not very busy places. I can remember at one exhibition, it was half-jokingly remarked that, having run the 10:48 train, we could all now go off and have a cup of tea and a biscuit, there being no more trains until 12:04! Of course, we can speed up the timetable so that trains run in quick succession, but this is not really good enough. We often require more variety than this and particularly we want to have the possibility of several things happening at once, for example some shunting going on in the yard whilst up and down passenger trains run through the station. This keeps up the interest of the operators, and, if the layout is at a show, the public as well.

In my own case, I have taken a bit of a liberty with the track plan to facilitate this. On the prototype, shunting took place up and down the main line. In the case of a down goods train, the train would run forward to the advanced starter signal, then shunt its wagons from here into the goods yard. (In the early days a 'shunt-ahead' signal was provided on the starter signal post for this purpose). In the case of an up goods, this would be backed over the crossover into the yard and shunted from there. In each case, it was the goods train

engine that did the shunting, and whilst this was in progress there was little or no possibility of a passenger train being able run through the station.

To allow shunting and through passenger trains simultaneously, on my model the very short length of line at the south end of the yard (on the original this was a trap point only) has been extended into a head shunt. The station has also been provided with a pilot engine (which never existed on the prototype), so that shunting operations can continue even when the goods train has left – wagons can be positioned for loading or unloading then marshalled ready to leave on the next goods train etc. Whilst this is going on, passenger trains can run up and down through the station, thus providing plenty of action.

When a goods train arrives, therefore, the plan is to have the goods engine back its train of wagons into the yard, then retire to the up platform where there was a water crane. The idea is that the engine is taking on water whilst the station pilot removes or adds whatever wagons are necessary to the train. However, this does mean that the up platform is blocked for the duration, so to keep traffic moving, a facility was added to allow the goods engine to be moved to the down platform and back, should an up passenger train be due before the goods is ready to leave.

Whatever your layout, the first step in planning signalling for it should be a process similar to the above of deciding what it is you want to achieve. Having decided on the general operating goals you can now follow a series of steps to come up with a detailed signalling plan.

Step 1 How Many Operators?

How many operators does your layout require? Basically, this is a question of the maximum number of simultaneous train movements you want at any time. You could have additional operators, however, doing background tasks like assembling trains in a fiddle yard. You will also, of course, need at least one signalman to operate the lever frame. In the case of Whitchurch, three 'train' operators are needed. One is based at the 'Newbury' (N) end of the layout and drives the 'up' trains. Another is based at the 'Southampton' (S) end and drives down trains, and a third is based in the goods yard (Y) to do the shunting. The N and S operators are also responsible for assembling trains in the fiddle yards at either end of the layout. A fourth operator works the lever frame.

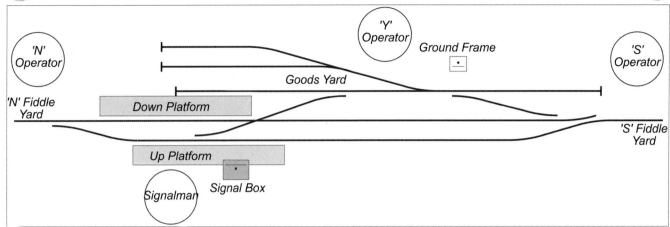

Figure 4.1 *- Operating positions.*

Step 2 What Train Movements?

The next step is to write down a list of all the train movements you might require, and which operator will be responsible for them. In the following, NFY and SFY stand for the N and S fiddle yards, UP and DP are the up and down platform roads, and GY the goods yard.

Passenger and Through Goods Trains

a. Train from NFY to DP (S drives).
b. Train from DP to SFY (S).
c. Train from SFY to UP (N).
d. Train from UP to NFY (N).

Pick-up Goods Trains

e. Train/Engine from UP to NFY (Y).
f. Train/Engine from UP to GY (Y).
g. Train/Engine from GY to UP (Y).
h. Train/Engine from DP to SFY (Y).
i. Train/Engine from SFY to GY (Y).
j. Train/Engine from GY to SFY (Y).
k. Train/Engine from SFY to DP (Y)
l. Train/Engine from SFY to UP (Y)
m. Shunting within the yard (Y)

Moving Engine between platforms

n. Engine from DP to NFY (N)
o. Engine from NFY to UP (N)
p. Engine from UP to NFY (N)
q. Engine from NFY to DP (N)
r. Engine from DP to SFY (S)
s. Engine from SFY to UP (S)
t. Engine from UP to SFY (S)
u. Engine from SFY to DP (S)

Some of these are pretty obvious, but some may need

explanation. When an up stopping goods train arrives, for example, the N operator drives it into the up platform (c). If necessary to clear the crossover points, the yard operator draws the train forward (e), then backs the train into the yard (f), uncouples the engine and moves it back to the up platform (g). When shunting is complete, the yard operator moves the engine back onto its train (f) couples up and pulls the train back into the up platform (g). Should an up passenger train arrive before shunting is finished, we do not want to interrupt the yard operator, and since it is the N operator who is being held up, we give him the job of moving the light engine to the down platform, move (n) then (o), and back.

In the case of a down stopping goods train, the S operator drives it into the platform (a). We can follow move (a) by either moves (h), (i), (j) and (l), in which the yard operator backs the train into the yard then places the engine in the down platform, or, if the yard operator is busy, the S operator can uncouple from the wagons in the down platform and move the engine to the down platform himself, moves (r) and (s). When the yard operator is ready, he can then pick up the wagons left in the down platform road, moves (j), (k), (h), and (i). When shunting is complete, the station pilot can place the wagons in the down platform, then return to get on with shunting, whilst operator S goes to pick up his train, moves (t) and (u), and drive it away (b).

Step 3 What Track Sections?

This step is applicable only if you are using old-fashioned DC control. If you are using DCC, and have decided not to break up your track into sections then you need not bother about this. Basically, the idea is to decide, for each move, what parts of the track the engine needs to run along, and to draw up a table indicating which operator's train controller is connected to which section of track for each move. Note that the head shunt (HS) in the goods yard is isolated from the rest of the yard (GY). This is so that we can isolate the

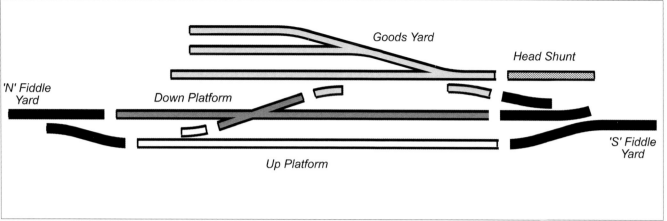

Figure 4.2 - *Track sections.*

	NFY	DP	UP	SFY	GY	HS
	(F3)	(F4)	(F5)	(F1)	(F6)	(F2)
Move						
a	S	S				
b		S		S		
c			N	N		
d	N		N			
e	Y		Y			
f			Y		Y	
g			Y		Y	
h		Y		Y		
i				Y	Y	
j				Y	Y	
k		Y		Y		
l			Y	Y		
m					Y	Y
n	N	N				
o	N		N			
p	N		N			
q	N	N				
r		S		S		
s			S	S		
t			S	S		
u		S		S		

Table 4.1

shunting engine in the head shunt whilst a goods train is backing into the yard from the up platform.

For example, move (a) is the arrival of a train from the N fiddle yard to the down platform. This will require the S operator's controller to be connected to the N fiddle yard and the down platform track sections, represented by NFY

and DP. Move (b) will require S's controller to be connected to DP and SFY, and so on. These connections are drawn up in the form of a table, as in table 4.1. F1, F2, … are the track feeds to each of the sections.

Step 4 What Signals?

If we were going for a control panel approach, using conventional cab control (as described in chapter 1), then we could simply place switches on each of the track sections, and use them to connect to the appropriate controller (N, S, or Y). However, this is not the way the prototype worked. We seek to use prototypical signalling in as realistic a way as possible, and to derive from this signalling system the connections required in table 4.1. This approach will involve some compromise and some creative use of the signalling possibilities.

Let us first consider the up and down passenger trains. These can be signalled in a fairly straightforward manner using almost exactly the same approach as the prototype. Our down home signal, for example, can simply make a connection between the S controller and NFY and DP. Our down starter connects DP and SFY to the same controller. Thus, the S operator can drive a train into the down platform when lever 1 is pulled, then from the platform into his fiddle yard when lever 2 is pulled. The up home and starter signals will perform a similar function in relation to the N operator.

We now move on to some of the shunting moves, and here we do need to employ one or two 'dodges' to achieve the desired result. A down goods train, as we have discussed, must be drawn forward into the single line section south of the station, move (h), before being backed into the yard, move (i), under the control of the Y operator. If we signal this move (h) with the down starter (lever 2), how do we distinguish it from move (b)? The latter is the same move, but under the control of the S operator. In this case, we can adopt a simple expedient – use a 'shunt-ahead' subsidiary signal on the down starter post. If the main signal is pulled off, we connect DP and SFY to S operator, if the subsidiary signal is pulled off, we connect the same sections to Y operator.

In the case of an up goods train, before the train can be backed over the crossover into the yard, the Y operator may require to pull it forward to clear the crossover points, move (e) in table 4.1. Now this, again, requires the same track sections as move (d), a departing up passenger train, except that the connections are to the Y operator rather than the N operator. We could employ the same logic here, and install a shunt-ahead signal on the up starter post. However, there is an alternative approach, which provides an elegant method of overcoming the problem, and saves levers as well.

We note that, in addition to the Y operator's requirements in shunting goods trains into and out of the yard, there are some shunting moves required of the N operator, namely the moving of a light engine from the up to the down platform to clear the way for an up passenger train, moves (p) and (q). We can signal these with ground discs, but how do we distinguish, for example, between (p), which requires operator N to be connected, and move (e) which requires operator Y? The answer lies in a little piece of lateral thinking.

If the Y operator is busy shunting the yard, he can't be doing anything else. To shunt the yard, he will require a ground frame. We can give this ground frame a 'king' lever, which releases all the other levers in the ground frame. Furthermore, we can make this 'king' lever electrically locked until a ground frame release lever is pulled on the main lever frame. We can then apply a rule that if the ground frame release lever is pulled, the Y operator is presumably busy shunting the yard, so the shunt signals on the up and down platforms refer to the N operator. If the ground frame release lever is normal, we assume the Y operator is doing the shunting on the main line. Similar logic applies at the other end of the loop to the S operator.

Finally we mention a couple of dodges that can save levers. At the time of writing, the lever frame kit I used costs in excess of £15 per lever, so making one lever do the job of several is an economically sensible idea. We note that moves (h) and (k) are identical in terms of the track connections required. The only difference is the direction in which the engine is driven, either from down platform to yard or the other way round. We could therefore signal the move with just one ground signal. Admittedly, it would be passed in the wrong direction on one move or the other, but ground signals are hardly noticed in practice, so you might feel that this is justifiable. There was a sort of precedent for this on LSWR lever frames where crossovers would sometimes be worked by a single 'push-pull' lever whose 'normal' position was in the half-way point of the frame. To select one direction through the crossover, you had to push the lever back, to select the other direction, you pulled it forward.

Another lever-saving idea, which was employed on the prototype, is to make a single lever control two or more different signals depending on the setting of points. For example, we have mentioned the need to shunt an engine from the up to the down platforms and back. We could have a single lever controlling two separate shunt signals, and determine which shunt signal comes off when the lever is pulled by the setting of the points at the north end of the loop. Indeed, if we apply both this dodge and the one in the previous paragraph, we can have a single lever do the work of four!

By means of the above techniques, we find that the layout can be adequately signalled by means of far fewer levers than the prototype's 27. A further significant saving can be made by dispensing with facing point lock levers. Very few modellers bother with these, and I have never seen a working one with a proper bolt that locks the points. In fact, my frame has just 15 levers, one of which is currently a spare, the layout having been rebuilt in such a way that it is no longer required. Figure 4.3 illustrates what the 14 operational levers do.

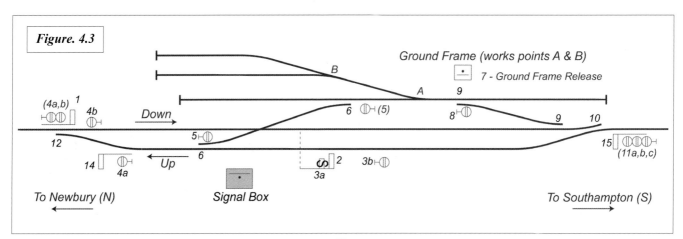

Figure. 4.3

Notice that lever 4 controls four ground discs (in fact, only two of these are modelled on my layout, the others being presumed to be outside the scenic limits of the layout). Lever 11 controls three ground discs (although these again are all outside the limits and so are not modelled). The way each type of train movement is to be signalled is indicated in table 4.2.

Observe also that each train movement involves a unique combination of lever pulls. In general, we should always pull points before signals (we shall see later how to ensure this by interlocking), and when restoring levers, we put signals back first. Where lever 7 is in parentheses, this is to indicate that 7 is already pulled (to release the ground frame) so pulling 7 is not technically part of the move. It is there simply to indicate that we want one of the fiddle yard operators (N or S) to drive, rather than the yard operator (Y); this is indicated by the asterisk (*) against signals 3, 4 and 11.

Signal(s)	Lever(s)	Movement	Track sections	Controller
1	1	Down train arrives	NFY,DP	S
2	10,2	Down train departs	DP,SFY	S
1,2	10,1,2	Down through train	NFY,DP,SFY	S
15	15	Up train arrives	SFY,UP	N
14	12,14	Up train departs	UP,NFY	N
15,14	12,15,14	Up through train	SFY,UP,NFY	N
3a	(not 7),10,3	Down shunt ahead	DP,SFY	Y
11c	10,9,11	Back train into yard	SFY,GY	Y
8	10,9,8	Engine/Train out of yard	GY,SFY	Y
11a	11	Back engine into up plat.	SFY,UP	Y
11b	10,11	Back engine into down plat	SFY,DP	Y
3b	3	Engine from up plat. to line (S)	UP,SFY	Y
4a	(not 7),12,4	Pull forward from up plat.	UP,NFY	Y
5	6,5	Up plat to/from yard	UP,GY	Y
4a*	(7),12,4	Up plat to/from line (N)	UP,NFY	N
4b*	(7),4	Down plat to/from line (N)	DP,NFY	N
3a*	(7),10,3	Down plat to line (S)	DP,SFY	S
3b*	(7),3	Up plat to line (S)	UP,SFY	S
11a*	(7),11	Line (S) to up plat	SFY,UP	S
11b*	(7),10,11	Line (S) to down plat	SFY,DP	S
	7	Release ground frame	GY,HS	Y

Table 4.2

operators (N or S) to drive, rather than the yard operator (Y); this is indicated by the asterisk (*) against signals 3, 4 and 11.

Step 5 Connect Track Sections

Each lever on my lever frame operates a microswitch which provides +12v when the lever is reversed, or -12v when the lever is normal. To connect the various track sections, we use relays. A relay is exactly like a switch, except that instead of turning it on and off with your finger, you operate it by supplying it with electrical current. The most convenient type of relays for our purposes are those that operate on 12v DC, and they come in a variety of configurations: single-pole, double-pole, four-pole etc. Each 'pole' of the relay has three associated terminals, or connections, see figure 4.4. A 'common' terminal (COM) makes contact with the 'normally closed' (NC) terminal when no current flows through the relay coil. At this time the third 'normally open' (NO) terminal is not connected to anything (upper diagram in figure 4.4). When 12v is connected to the operating coil of the relay (lower diagram), a magnetic field is set up which causes the 'common' contact to flip over from the 'normally closed' position to make contact with the 'normally open' terminal. Shown on figure 4.4 across the relay coil is a diode. This is there to protect against 'reverse voltage spikes', about which we will say little except that they can damage certain components. It is good practice to wire such a diode across each relay coil

as shown. However, to save cluttering up later diagrams we will not include them, restricting ourselves to those diodes that actually form a part of the logic of the circuit.

Each track section has one relay for each controller that we might require to connect to it. Thus, for example, track section UP (the up platform) will have three track relays, relay UPN will connect it to the N controller, relay UPY will connect it to the Y controller and UPS to the S controller. For each signalled move, we take a diode from the appropriate signal microswitch to the appropriate track relay, as shown in figure 4.5, which illustrates how

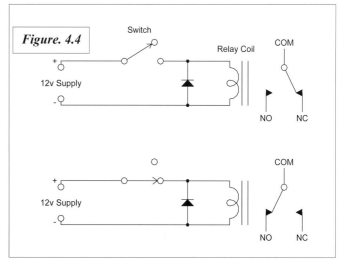

Figure. 4.4

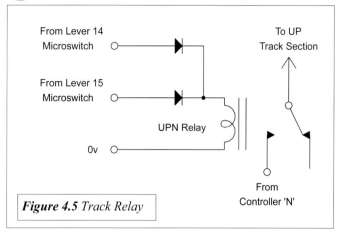

Figure 4.5 *Track Relay*

reversing either lever 14 or lever 15 will cause the UP track section to be connected to the N controller. This is just a snippet of a large array of relays. For those interested in seeing the complete circuitry, a full circuit diagram appears in Appendix 3.

In the next chapter, we shall take this layout design further, by adding suitable interlocking. Meanwhile, though, we shall conclude this chapter by looking at a more ambitious signalling project. This is a part of a club layout and is included to show just how much you can achieve using a lever frame only.

Itchen Bottom

This is a fictitious station, part of a fairly extensive club layout which in all boasts five lever frames and eight train controllers. It can therefore provide entertainment for a good number of operators, which is, after all, the main goal of a club layout. As you can see from the track diagram (figure 4.6), the station forms part of an oval of double track. It is controlled from a main lever frame of 22 levers. A smaller frame of just 10 levers controls a set of storage sidings. At the other side of the oval is a junction with some sidings and its own lever frame. From this junction, trains run into another section of the layout, which comprises a terminus station with a branch line going off from it. Two lever frames, one for the main line and one for the branch, control this part of the layout, along with two small ground frames for shunting.

With this many lever frames, there could be problems if they all had levers numbered from 1 upwards. Not only would this cause confusion for operators, but it would make a nightmare of trying to diagnose wiring problems. Therefore, it was decided that the first lever frame would be numbered with levers 1 to 20, the second would be numbered from 21 to 40, and so on. The Itchen Bottom (I.B.) station's lever frame is numbered from 61 to 82. The fact that there is no other lever on any of the other frames with a number in this range means that it is perfect clear

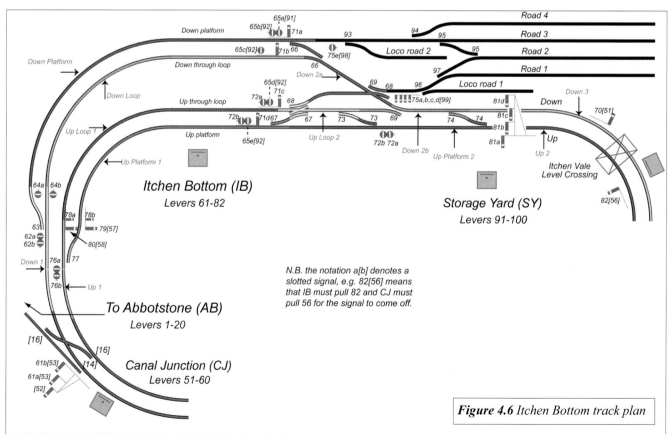

Figure 4.6 *Itchen Bottom track plan*

what is meant when you say "lever 71"!

We shall not repeat the whole process of describing the individual steps 1 to 5, as above, but instead will cut straight to the signalling diagram itself. In fact, the layout was not designed in the recommended manner. A lever frame, beautifully scratchbuilt in brass by one of the club's late members, was available and this happened to have 22 levers. It was not, therefore, a case of deciding how many levers we would need, but rather how to signal the existing station layout using exactly 22 levers!

If you count them up on the track diagram, there are 34 signals and 8 sets of points associated with the I.B. lever frame. To control all of these with just 22 levers was thus quite a challenge, and some levers control up to five different signals, depending on point settings. Take lever 61, for example. This controls the two stop signals on the left hand side numbered 61a and 61b. These are the home signals for, respectively, the down platform and the down loop. Which signal comes off when lever 61 is reversed depends on the setting of lever 63. If 63 is reversed, i.e. the points are set for the down platform, then pulling lever 61 will operate signal 61a. If 63 is normal, points are set for the down loop, then pulling 61 will operate signal 61b.

Similar logic has been applied to virtually all of the signal levers. Lever 65, for example, controls no less than five separate ground signals, depending on the setting of points 66, 67, 68 and 69. Table 4.3 lists all the signals and their functions.

How do we manage to operate this large number of signals, not to mention the points as well, with just 22 levers? The answer is, by using relays, as the next section explains.

Relay Logic

We can see how easy it is to implement the logic required for signals 61a and 61b above. We connect the output from lever 61, which will be +12v when the lever is pulled, to the common terminal of a relay whose operating coil is connected to lever 63. When 61 is pulled, but 63 is not, then 'common' (+12v from lever 61) will be connected to the 'normally closed' terminal, thus providing the signal we require for S61b. When 61 and 63 are both pulled, the +12v will be connected to the 'normally open' terminal, so providing the signal required for S61a.

This is a very simple piece of logic. We can write it down as a logic rule, using the symbol '&' to mean 'and', and the symbol '-' to mean 'not'. We write

S61a = L61&L63
S61b = L61&-L63

Signal	Function
S61a	Down 1 to Down Platform Home
S61b	Down 1 to Down Loop Home
S62a	Down 1 to Down Platform run round
S62b	Down 1 to Down Loop run round
S64a	Down Platform to Down 1 shunt
S64b	Down Loop to Down 1 shunt
S65a	Down Platform to Top Storage
S65b	Down Platform to Bottom Storage
S65c	Down Loop to Bottom Storage
S65d	Up Loop 1 to Bottom Storage
S65e	Up Platform 1 to Bottom Storage
S70	Down 3 Advanced Starter
S71a	Down Platform to Down 3 Starter
S71b	Down Loop to Down 3 Starter
S71c	Up Loop 1 to Down 3 Starter
S71d	Up Platform 1 to Down 3 Starter
S72a	Up Loop 1 run round
S72b	Up Platform 1 run round
S75a	Up Platform 1 to Bottom Storage
S75b	Up Loop 1 to Bottom Storage
S75c	Down Loop to Bottom Storage
S75d	Down Platform to Bottom Storage
S75e	Down Platform to Top Storage
S76a	Up 1 to Up Loop 1 run round
S76b	Up 1 to Up Platform 1 run round
S78a	Up Loop 1 to Up 1 shunt
S78b	Up Platform 1 to Up 1 shunt
S79	Up Platform 1 to Up 1 Starter
S80	Up Loop 1 to Up 1 Starter
S81a	Up 2 to Up Platform 1 Home
S81b	Up 2 to Up Loop 1 Home
S81c	Up 2 to Down Loop Home
S81d	Up 2 to Down Platform Home
S82	Up 2 Outer Home

Table 4.3

to mean, S61a is lever 61 and lever 63, whereas S61b is lever 61 and not lever 63. The position of all other levers is irrelevant to this rule. An example of a more complicated rule is

S65c = L65&L69&-L66&-L68

which means S65c comes off when levers 65 and 69 are both reversed, and 66 and 68 are both normal. Table 4.4 gives a complete breakdown of the logic which enables all 34 signals to be controlled:

Looking at table 4.4, we see that lever 65 involves a fairly complicated piece of selection logic. Depending on the setting of points levers 66-69, pulling lever 65 could operate one of five different ground discs to signal a move from either the down platform to the top storage yard, or from the down platform, down loop, up loop or up platform into the bottom storage yard. To accomplish this, we use four *logic relays*, each operated by one of the points levers, as in figure 4.7.

Signal	Lever(s)
61a	L61&-L14&L63
61b	L61&-L14&-L63
62a	L62&L63
62b	L62&-L63
64a	L64&L63
64b	L64&-L63
65a	L65&-L69&-L66
65b	L65&L69&L66
65c	L65&L69&-L66&-L68
65d	L65&L68&-L67
65e	L65&L68&L67
70	L70
71a	L71&-L70&-L69&L66
71b	L71&-L70&-L69&-L66
71c	L71&-L70&L69&-L67
71d	L71&-L70&L69&L67
71a*	L71&L70&-L69&L66
71b*	L71&L70&-L69&-L66
71c*	L71&L70&L69&-L67
71d*	L71&L70&L69&L67
72a	L72&L73
72b	L72&-L73
75a	L75&L68&L67
75b	L75&L68&-L67
75c	L75&L69&-L66&-L68
75d	L75&L69&L66
75e	L75&-L69&-L66
76a	L76&-L77
76b	L76&L77
78a	L78&-L77
78b	L78&L77
79	L79
80	L80
81a	L81&-L74&-L73
81b	L81&-L74&L73
81c	L81&L74&-L66
81d	L81&L74&L66
82	L82

Table 4.4

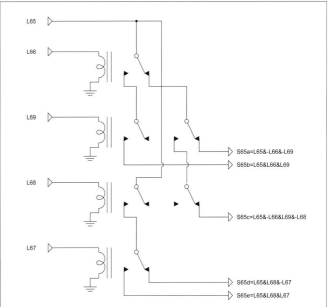

Figure 4.7 Logic relays

Similar logic applies to all of the other signals, except those such as S70 and S80, where a lever works only a single signal. In the case of lever 71, notice that there are four separate down starter signals, S71a, S71b, S71c and S71d, to correspond with the two platforms and two loops. However, in Table 4.4 we have shown eight separate logic rules, labelled S71a, S71b,…etc. and S71a*, S71b*,…etc. The reason for this is that different controllers are involved, as was the case with the previous layout. We shall see below the significance of this.

Connecting Track Sections

As can be seen from the track diagram, I.B. has some 14 track sections, as follows:

Section Name	Abbreviation
Down 1	Dn1
Down Platform	DnPt
Down Loop	DnLp
Down 2(a)	Dn2a
Down 2(b)	Dn2b
Down 3	Dn3
Up 1	Up1
Up Loop 1	UpLp1
Up Loop 2	UpLp2
Up Platform 1	UpPt1
Up Platform 2	UpPt2
Up 2	Up2
Top Storage Roads	TopSt
Bottom Storage Roads	BtmSt

There are four train controllers, colour-coded as follows: I.B. 'Up' controller (Blue), 'Down/Storage' controller

(Yellow), Canal Junction 'Down' controller (Green) and 'Up' controller (White). The general rule we adopt is that each controller drives trains towards himself, the idea being that it is harder to stop in the right place the further the train is from you. Thus a move from the C.J. down line into the I.B. down platform, signalled by S61a (see table 4.3), will be under the control of the yellow controller (Y). This will require the track sections labelled Dn1 and DnPt to be connected to controller Y. For each of the moves listed in

Signal	Connect Track Section(s)...	to...
61a	Dn1,DnPt	Y
61b	Dn1,DnLp	Y
62a	Dn1,DnPt	B
62b	Dn1,DnLp	B
64a	Dn1,DnPt	B
64b	Dn1,DnLp	B
65a	DnPt,TopSt	Y
65b	DnPt,Dn2a,BtmSt	Y
65c	DnLp,Dn2a,BtmSt	Y
65d	UpLp1,UpLp2,Dn2a,BtmSt	Y
65e	UpPt1,UpLp2,Dn2a,BtmSt	Y
70	Dn3	G
71a	DnPt,Dn2a,Dn2b,Dn3	Y
71b	DnLp,Dn2a,Dn2b,Dn3	Y
71c	UpLp1,UpLp2,Dn2b,Dn3	Y
71d	UpPt1,UpLp2,Dn2b,Dn3	Y
71a*	DnPt,Dn2a,Dn2b,Dn3	G
71b*	DnLp,Dn2a,Dn2b,Dn3	G
71c*	UpLp1,UpLp2,Dn2b,Dn3	G
71d*	UpPt1,UpLp2,Dn2b,Dn3	G
72a	UpLp1,UpLp2,UpPt2	Y
72b	UpPt1,UpPt2	Y
75a	UpLp1,UpLp2,Dn2a,BtmSt	Y
75b	UpLp1,UpLp2,Dn2a,BtmSt	Y
75c	DnLp,Dn2a,BtmSt	Y
75d	DnPt,Dn2a,BtmSt	Y
75e	DnPt,TopSt	Y
76a	Up1,UpLp1	B
76b	Up1,UpPt1	B
78a	Up1,UpLp1	B
78b	Up1,UpPt1	B
79	Up1,UpPt1	W
80	Up1,UpLp1	W
81a	UpPt1,UpPt2,Up2	B
81b	UpLp1,UpLp2,UpPt2,Up2	B
81c	DnLp,Dn2a,Dn2b,Up2	B
81d	DnPt,Dn2a,Dn2b,Up2	B
82	Up2	B

Table 4.5

table 4.3, we now work out which controller is to be used, and which of the track sections are to be connected to that controller. The result is shown in table 4.5.

From the above table, we can construct an array of *track relays*, each of which, when activated, connects a particular track section to a particular controller. For example, we want to connect the track section DnPt to the 'Yellow' controller when any one of S61a, S65a, S65b, S71a, S75d or S75e is activated. We therefore take each of these signals via a diode to the coil of a track relay, whose 'common' terminal is connected to controller Y, and whose 'normally open' terminal is connected to the live feed on the down platform, DnPt. See figure 4.8.

Observe in the figure that a yellow light emitting diode (LED) is placed in parallel with the relay coil. This LED (the 12v variety), along with others like it, can be used to provide an illuminated display panel, showing which track sections are connected to which controller. This is an extremely useful accessory, especially if certain signals (like those mentioned earlier) are beyond the scenic limits of the layout and are therefore not modelled. The operators can still see when they are cleared by the illumination of the appropriate route on the display panel.

We now see why there are two signals S71a, which we have called S71a and S71a*. Both of these cause the same semaphore arm to come 'off', namely the down platform starter. If lever 71 is pulled without lever 70 (the down advanced starter), then this move is to be made from the down platform only as far as the down advanced starter signal. Here the engine must wait. This move is made under the control of the I.B. down controller (Y). However, if 71 is pulled in conjunction with lever 70, then the move is signalled all the way through from the down platform to the junction. In this case, we want the C.J. down controller (G) to drive. Hence, S71a is one of the signals that activate the DnPt Y track relay (shown in figure 4.8), whereas S71a*

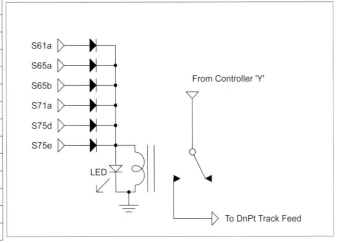

Figure 4.8 *Track Relay*

activates the DnPt G track relay.

Offering and Accepting Trains

Some modellers favour building miniature replicas of block telegraph instruments and using these to 'offer' and 'accept' trains between operators, using the correct bell codes etc. This, of course, is the truly authentic way to do things, and figure 1.5 illustrates just such a scheme. However, it does not appeal to everybody, particularly in the case of a layout destined to be shown to the public. Holding up the action whilst a long sequence of ting-ting-ting sounds is exchanged can leave the general public somewhat bemused, fascinating though it may be to the aficionado. Having had no personal experience of such an approach, I cannot comment.

Our preference in our local club is to use the signals. We have seen in the previous section how lever 70 is the advanced down starter for I.B. When this lever is pulled by the I.B. signalman, we arrange that it causes a flashing LED to illuminate on the C.J. display panel. This means 'a train is ready for you to drive'. The C.J. operator can now accept this train (assuming his block section is clear to the appropriate clearing point), which he does by pulling his down home signal. When he does this, the LED stops flashing. A further refinement is to have the advanced starter slotted (like the slotted distant signals described in the previous chapter). In other words, the arm on S70 drops only when both the I.B. operator has pulled his up advanced starter signal lever, and the C.J. operator has pulled his down home. This gives a visual indication that the train has been offered and accepted.

It may be the case that one of the operators does not have signals operated by levers, for example if the operator is running a set of hidden storage sidings or a fiddle yard. In this case, a simple push-button switch can be used instead. This will set off a flashing LED which continues until the other operator pulls the appropriate lever to accept the train. This is the approach used in the 'Whitchurch' layout.

The Complete Works

For those interested in the nitty-gritty of these things, Appendices 3 and 4 contain complete circuit diagrams of the 'Enigma' boxes for both Whitchurch and Itchen Bottom. The term 'Enigma Box' was coined by John Shaw, who built our club's first one. It shares its name with the wartime encryption/decryption machine. However, we are not certain whether it is named after this or whether its name refers to the fact that no-one can puzzle out how it works! The name has stuck.

Should you decide to build your own Enigma Box these descriptions may provide useful examples. It is important to remember, whilst designing and constructing your equipment, that you will at some point in the future have to maintain it. Wires will break, solder joints will fall apart, relays will fail and miscellaneous bits of extraneous matter will fall into the box causing short circuits. All of these problems will need to be remedied, and it is far easier to do this if the components are laid out in a logical order and the wiring is kept tidy.

Extensive use is made of relays, both for implementing the logic required for signalling and for connecting controllers to track. It is advisable to group these together by function (logic relays, track relays) and to label them clearly for ease of maintenance. It will also be found easier if the plug-in variety are used, rather than those that are soldered into a circuit board. A useful type of plug-in relay socket is that designed for either surface mounting, or mounting on 35mm DIN rail (DIN 46277 or BS 5584). This has excellent accessibility, enabling you to plug and unplug the relay, also to connect and disconnect wiring using the screw terminals without having to do any dismantling or unsoldering.

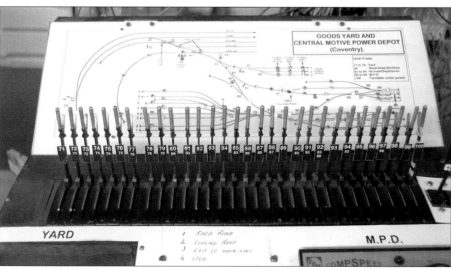

Figure 4.9 – *This entire complex layout is controlled solely by the levers. There are NO SWITCHES!*

5

BUILDING A LOCKING LEVER FRAME

In the previous chapter, we saw how a complete layout can be controlled entirely from one or more lever frames. In essence, we have used the lever that operates the signal to connect train controllers to the appropriate track section, in much the same way as would be the case on the prototype if we used the absolute block system and employed automatic train control as described in chapter 2. However, there is one important thing we need to add, and that is interlocking. As things stand, we could pull two levers which connected two different controllers to the same track section. This would not necessarily be a problem with DCC, but with conventional DC control it would be highly undesirable and could damage the controllers.

In this chapter, therefore, we shall see how to build a locking lever frame, which not only accomplishes the tasks required on the prototype (points set before signals released, no conflicting signals etc. as described in chapter 2), but also prevents electrical problems such as multiple controllers being connected to the same track.

Levers

Like semaphore signals, prototype levers differed from company to company. A popular type was that made by Stevens & Co, a model of which is illustrated in figure 5.1. The lever is pivoted at its lower end, and runs in a slot in a so-called quadrant plate. Alongside the slot for each lever, the quadrant plate has a raised portion which does not quite reach the extremities of the slot. Attached to each lever is a *catch dog* which is able to slide a little way up and down the lever. At the extremities of the slot, the catch dog drops onto the quadrant plate itself, and so prevents the lever from moving. To free it, the signalman has to close the *catch handle*, which lifts the dog clear of the raised portion of the quadrant plate thus allowing the lever to move. The purpose of this mechanism is to prevent the lever from moving by itself, which, if it were mechanically lifting a signal arm, say, it would tend to do.

In miniature, you need to have a catch dog and lifting mechanism which move extremely smoothly if you want to scratch-build a lever along the lines of the prototype. This is because the catch dog is extremely light, and relying on gravity alone to cause it to drop means that the slightest hint of friction will prevent this from happening. You can add a tension spring to pull it down onto the quadrant plate if you wish.

To drive the logic relays and track relays, as described in the

previous chapter, we need to derive an electrical signal from each lever. The easiest way to do this is to fit a microswitch beneath each lever. You can have the levers supply a voltage signal when pulled, and nothing whilst in the normal position. Alternatively, you can have them supply a positive voltage when pulled, and a negative one whilst normal. The advantage of the latter is that the voltage can then be used directly for operating point motors, but it does mean that diodes will have to be used for operating relays, since a relay will operate just as well on -12v as +12v.

If you do not wish to scratch build your lever frame, there are a number of kits on the market which provide you with the necessary parts. Suppliers of these are listed in Appendix 1.

Lever Colours

On a prototype lever frame the levers will be painted in different colours according to their function. This was invariably the case with full size levers, but frequently with miniature levers in power frames also. The aim was to

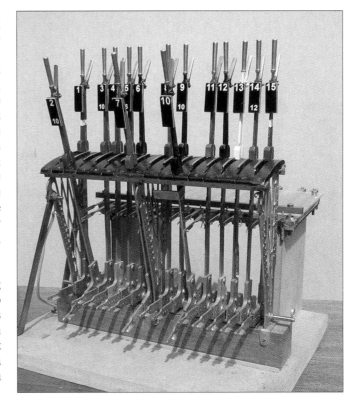

Figure 5.1 - *One-twelfth scale model of Stevens & Co. lever frame.*

assist the signalman in quickly recognizing which lever was which. Before long, it becomes easy to pick out any given lever from its position in relation to the pattern of colours along the frame, without the need to read any numbered labels.

Once again, it must be emphasised that practice varied from railway to railway, especially in the early years. The following is therefore a general guide only.

Stop signals, including subsidiary signals, were operated by red levers. In cases where the signal was released by 'line clear' from the box in advance, the red lever would have a broad white stripe about half-way up. At the time when distant signal arms were painted exactly like stop arms, their levers would be green, a reminder of the early colour code for lamps when red meant danger, green caution and white clear. However, once distant signals began to be painted yellow, their levers followed suit. Sometimes, a stop signal and its distant would be worked from the same lever, in which case the lever would be red for its lower half and yellow for its upper.

Points levers were black. If the points were facing, they would invariably be equipped with a facing point lock, and this would have a blue lever. On many smaller lines, so-called 'economical facing point locks' would be employed. These operated both the point and the lock from the same lever, which would be painted black on its lower half, and blue on its upper.

Release levers for such things as level crossing gates and turntables were generally brown, as were mechanical ground frame release levers. Where a ground frame had an electrical release the lever was brown (lower half) and blue (upper), although sometimes green was used for this. 'King' levers, used to unlock a whole series of other levers, were also painted green.

Detonator placer levers were white with black chevrons pointing upwards for the 'up' line and downwards for the 'down'.

Levers which were not used, either because the frame was a standard one built with a fixed number of levers, not all of which were required, or because a particular siding or signal had been taken out of use, were painted plain white.

Levers were numbered from left to right and each carried a number plate. Often, where a lever was released by one or more other levers, the plate would also carry the number(s) of the lever(s) that should be pulled before this lever. In figure 1.5, for example, lever 14 number plate carries a small '12' below, meaning that 12 releases 14 and so must be pulled first.

One piece of signalman's jargon we shall use frequently is that a lever is described as being 'normal' or 'reverse'. If a lever is normal, it is in the position furthest from the signalman, and the signal it controls will be at danger. If the lever controls a set of points, the points will be set for their 'normal' route, i.e. the one shown on the signalbox diagram. If a lever is reversed, i.e. the signalman has pulled it towards himself, then the signal will be set to clear, or the points will be set to their alternate position.

Principles of Interlocking

We come now to the nub of the matter. How do we ensure that conflicting levers cannot be pulled? The answer is, by interlocking, and we hinted in chapter 2 at how some of the early interlocking mechanisms worked. There were various other schemes, some of which interfered with the release of the catch handle or dog, others obstructed the slot in the quadrant plate. However, by far the most popular was the system of *tappet locking*, originated by the firm of Stevens & Co. This is by far the easiest to implement in a small scale, so it is the one we shall focus on here.

Consider the following very simple layout (figure 5.2). A train may arrive from the left (signal 1) and enter either of two platforms, depending on the setting of points 2. Signals 3 and 4 allow a train in either platform to set off, again depending on points 2. When points 2 are normal, the train may set off from platform A. When they are reversed, a train in platform B may set off.

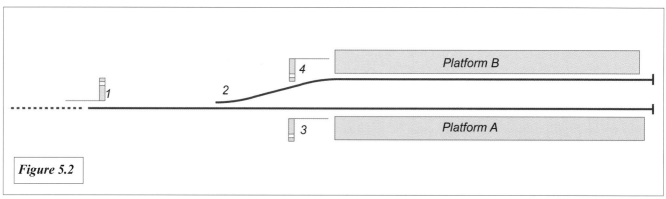

Figure 5.2

We wish, therefore, to implement a set of rules which will ensure that lever 1, when pulled, prevents the pulling of levers 3 and 4, we say '1 *locks* 3 and 4'. We also wish to ensure that lever 3 can only be pulled if 2 is normal, so '2 *locks* 3', and that lever 4 can only be pulled if lever 2 is reversed, so we say '2 *releases* 4'. Finally, we have seen earlier that it is highly desirable to set points before we set signals, otherwise the points might be changed whilst a train is running over them. In the case of signals 3 and 4, we cannot change points 2 under the rules already stated. However, these rules do not prevent points 2 being changed whilst signal 1 is 'off'. We therefore add a further rule that says '1 *locks* 2 *both ways*'. That is, once lever 1 is pulled, it locks lever 2 in whatever position it is at that time. Lever 2 cannot be changed until 1 is put back normal.

To accomplish this, we attach to each lever a *tappet blade*, or *tappet*. This is a piece of flat bar which is free to slide backwards and forwards within a *locking tray* as the lever moves to and fro. In figure 5.3 we have four tappets, numbered 1-4 corresponding to levers 1-4 in the frame. When a lever is pulled, it will cause its corresponding tappet to slide forwards (downwards in the diagram). Also within the locking tray, and at right angles to the tappets, are a series of *locking bars*. In figure 5.3, we have one bar, A. It is by interfering with the motion of the tappets that these locking bars prevent (or permit) the movement of other levers in the frame. This is done by means of *locking dogs*, or *nibs*. These are most easily represented in a small scale by the heads of 6 or 8BA screws in threaded holes in the locking bar, and these are shown as dark circles in figure 5.3. The dogs engage with notches filed into the sides of the tappets, so that as the tappet moves, the dog is pushed out of the notch, causing the locking bar to move sideways.

In figure 5.3, we see that locking bar A has a dog located in the notch in tappet 1. There are two further dogs on locking bar A, adjacent to tappets 3 and 4. However these are alongside their corresponding notches, not within them. If lever 1 is pulled into its reverse position, tappet 1 will move downwards (in the diagram) thus pushing the dog out of its notch, causing locking bar A to move to the right. Tappets 3 and 4 are therefore no longer free to move, since the two dogs alongside them are now pushed into their notches, and so long as 1 remains reversed, levers 3 and 4 are locked.

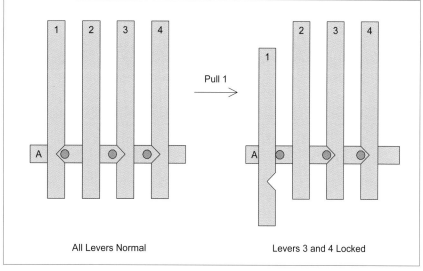

All Levers Normal — Levers 3 and 4 Locked

Above - Figure 5.3.   Below - Figure 5.4

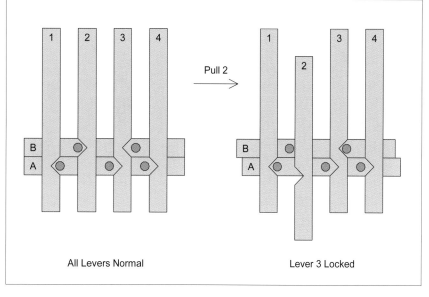

All Levers Normal — Lever 3 Locked

Notice also from the above diagram that if either lever 3 or 4 were to be reversed first then locking bar A would be unable to move to the right, thereby locking lever 1. Thus we see that lockings are always mutual: if lever *a* locks lever *b*, then it follows that lever *b* will lock lever *a*. We next turn to our second required locking rule, namely that lever 2 locks lever 3. This is accomplished with locking bar B in figure 5.4.

With all levers normal, lever 3 is free to move. However, if we pull lever 2, this will force locking bar B to the left, placing its right hand dog in the notch on tappet 3, thereby locking it in position. Notice that we have chosen this time to move the locking bar left instead of right, both are equally permissible and in some cases (though not this one, as it happens) it may be necessary to allow clearance. Our third rule is a 'release' rule. We want lever 2 to release lever 4.

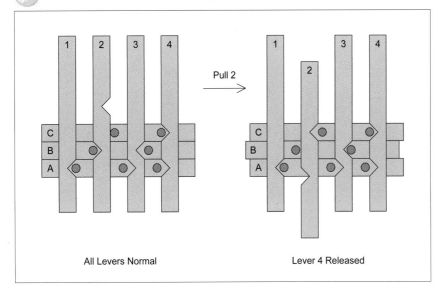

All Levers Normal   Lever 4 Released

Above - Figure 5.5.   Below - Figure 5.6

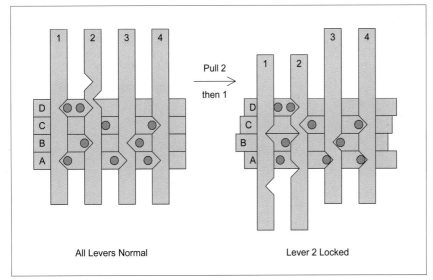

All Levers Normal   Lever 2 Locked

free to be pulled. Thus, 2 has released 4, as required. Notice that if lever 4 were now pulled, locking bar C would move left, and this would lock lever 2 in its reversed position. This is a general rule : if *a* releases *b*, then *b* locks *a* in reverse. Finally, we turn our attention to the 'both ways' lock.

In figure 5.6, with all levers normal, 2 is free to move. However, if 1 were to be pulled, it would move locking bar D to the right, locking 2 in its normal position. If, on the other hand, 2 were to be pulled first, and then 1 were pulled, this would cause 2 to be locked in reverse, as shown. Thus with 1 normal, 2 is free to move either way. However, when 1 is pulled, 2 is then locked in whichever position it happens to be at that time, hence 1 locks 2 both ways. Notice that a 'both ways' lock, unlike an ordinary lock, is *not* mutual. 1 locks 2 both ways, but the movement of 1 is unrestricted by 2.

Conditional Locks

On occasions, we may need to make a lock between two levers conditional on the setting of a third lever. For a very simple example, consider the track plan in figure 5.7, which is intended as a small fragment of a larger layout. Lever 5 signals a move from A to C or D, depending on the setting of the points 6, and lever 7 signals a move from D to A or B, again depending on the setting of points 6.

In figure 5.5, we see that, with all levers normal lever 4 is locked, because in order for tappet 4 to move, locking bar C must move left. However, it cannot do so, because there is no notch in tappet 2 alongside the left-hand dog in locking bar C. When lever 2 is pulled, however, this brings the upper notch in tappet 2 into alignment with the dog. Therefore, locking bar C can now move left, and lever 4 is

When points 6 are normal, there is no reason why levers 5 and 7 should not be pulled together. However, when 6 is reversed, this could give rise to a head-on collision. Thus we wish to have 5 lock 7 when 6 is reversed. Referring to figure 5.8, to arrange this, we employ a locking bar which is divided into two sections (shown as E and F) with ends raised above the level of the tappets, and a short gap between them. This means that the two sections can move

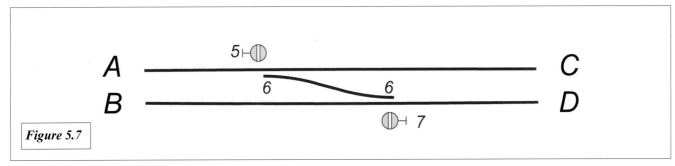

Figure 5.7

independently, so the two locking dogs do not cause any locking between levers 5 and 7, as in the left-hand diagram of figure 5.8.

However, when 6 is pulled, a short bar, pivotted so it can swing left to right, is placed between the two raised ends of bars E and F, as shown in the centre diagram. Now, if 5 is pulled, the motion of locking bar section E is transmitted via the swinging bar to section F, thus locking lever 7. So we have the desired result: 5 locks 7 when 6 is reversed.

Sequential Locking

Finally, we turn our attention to the implementation of sequential locking. We saw in chapter 2 that a key feature of the Sykes 'lock and block' procedure was that a section signal, having been cleared to allow a train to enter a section, and then restored to danger behind the train, could not then be cleared a second time until the home signal at the end of the section had itself been restored to danger after the train had passed out of the section. Not only was this a key safety procedure on the prototype, but it can be a very useful device on a model to protect our trains and controllers. Consider the snippet of line depicted in figure 5.9.

This represents a length of line with three track sections. The breaks between these are marked in the conventional manner. Suppose we have three operators, X, Y and Z. Operator X offers a train to operator Y, who accepts it. Signal 1 is cleared, connecting the left-hand and middle track sections to Y's controller. However, Y could offer the train on to Z, making it a through train. If we are following the principle that operators always drive trains towards themselves, then pulling both 1 and 2 will connect all three sections to Z. Once the train is out of the left hand section, signal 1 can be restored to danger. However, suppose X now offers another train to Y, who accepts it, signal 1 is cleared and Y thinks he is driving this second train. Not if 2 is still clear, he isn't! Having both signals clear will again connect all three sections to Z, who is therefore driving both the first train (that he accepted) and the second one (about which he knows nothing!).

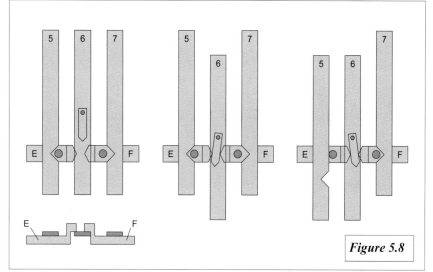

Figure 5.8

The problem arises because signal 1 has been cleared before signal 2 has been put back to danger. Sequential locking will ensure that lever 1 can only be pulled if lever 2 is normal. This, in turn, will mean that signal 1 can only be cleared once signal 2 is restored to danger. This is often described by saying '2 normal releases 1'. One way to implement sequential locking is illustrated in figure 5.10.

On the left, we see that tappet 1 has a square, rather than a triangular notch, and that the dog in locking bar A (shown shaded), instead of being round (i.e. the head of a set screw) is also square. The spring is pushing A to the right. However, it is unable to move as the right hand dog has no corresponding notch in tappet 2. In the centre, lever 1 has been pulled, followed by lever 2. The spring is still pushing A to the right, however it is now prevented from moving right by tappet 1, the 'square' notch no longer being aligned with the dog. However, on the right hand side of the diagram, lever 1 has been restored normal. Now the square dog is pushed by the spring into the notch in tappet 1. Because of the shape of the dog and the notch, locking bar A cannot be pushed back to the left by lever 1. Thus, lever 1 is now locked normal for as long as 2 remains reversed. Setting 2 back to normal will push locking bar A to the left, and restore the original configuration.

Adjacent Lever Release

It will sometimes happen, with cross-overs for example, that a lever will be required to release an adjacent lever. In this

Figure 5.9

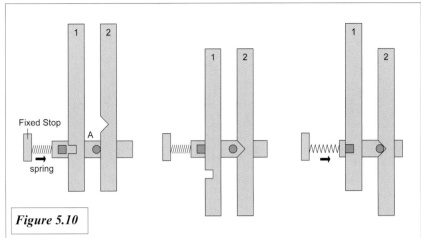

Figure 5.10

draw up our locking chart, we shall have need of the track diagram (see figure 5.11). We shall also need the table of signals and the track section/controller connections we require them to make, so this is repeated here to save the reader having to constantly refer back to the last chapter. Note also that we have used the same nomenclature as in *Table 4.2* to indicate which lever pulls activate which signal, that is the '&' symbol for 'and' and '-' for 'not'.

case, rather than do things using tappets, locking bars and dogs, there is a much quicker and easier way to achieve the same result. Simply solder a brass bar onto the lever doing the releasing, below quadrant plate level, such that this prevents the motion of the adjacent lever. If you look carefully at lever 10 in figure 5.1, you will see such a bar, which ensures that 10 releases 9.

The Locking Chart

In these very simple examples, we have seen how a series of locking requirements can be converted into a mechanical arrangement of tappets and locking bars. We have done this one step at a time, making one locking bar carry out each locking rule. However, in a real situation, things are considerably more complicated. We keep track of all the locks and releases needed in a locking chart, which we shall now derive.

We saw in the previous chapter how 'Whitchurch' was signalled using 15 levers (1 of which is in fact a spare, modifications to the layout having made it redundant). How do we go from here to a complete locking chart for this layout? Furthermore, how do we then go from a locking chart to a mechanical design for a set of locking bars? To

We now make a list of the various locks and releases we require. We start by looking at those situations where a route has to be set through the points before a running signal can be pulled off. We start at the top of the table 5.1, and note that points 12 should be normal if 1 is to be reversed. Therefore, we require that 1 locks 12 (and, therefore, that 12 locks 1). By contrast, if we wish to pull 2, then points 10 should be reversed first. This will be accomplished by including a rule that 10 releases 2. In a similar manner, for trains in the other direction, we note that 10 should lock 15, and 12 should release 14. We also note that points 6 should be set normal for a train to enter or leave the up platform, so 6 should lock both 14 and 15.

With regard to shunting moves, we note that 6 is required to release 5, and both 9 and 10 should be reversed before 8 is pulled. This latter is accomplished if we say that 10 releases 9 and 9 releases 8. Where we have a situation in which a single lever (3, say) can operate one of two signals (3a, or 3b) depending on the setting of points (10), then obviously we cannot have any route-locking rules for this.

Having ensured that we have covered all the situations where a route must be set correctly before a signal can be pulled, we now turn our attention to a second category of locks, those connected with *conflicting signals*. Because our running signals are associated with points which lock or release them, it is generally unnecessary to worry about these. There is no purpose, for example, in 1 locking 14,

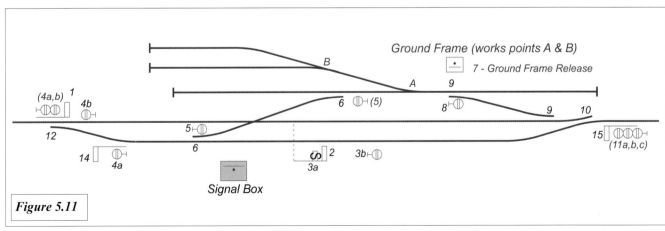

Figure 5.11

Signal (s)	Lever(s)	Movement	Track sections	Controller
1	1	Down train arrives	NFY,DP	S
2	10&2	Down train departs	DP,SFY	S
1,2	10&1&2	Down through train	NFY,DP,SFY	S
15	15	Up train arrives	SFY,UP	N
14	12&14	Up train departs	UP,NFY	N
15,14	12&15&14	Up through train	SFY,UP,NFY	N
3a	-7&10&3	Down shunt ahead	DP,SFY	Y
11c	10&9&11	Back train into yard	SFY,GY	Y
8	10&9&8	Engine/Train out of yard	GY,SFY	Y
11a	11	Back engine into up plat.	SFY,UP	Y
11b	10&11	Back engine into down plat	SFY,DP	Y
3b	3	Engine from up plat. to line (S)	UP,SFY	Y
4a	-7&12&4	Pull forward from up plat.	UP,NFY	Y
5	6&5	Up plat to/from yard	UP,GY	Y
4a*	7&12&4	Up plat to/from line (N)	UP,NFY	N
4b*	7&4	Down plat to/from line (N)	DP,NFY	N
3a*	7&10&3	Down plat to line (S)	DP,SFY	S
3b*	7&3	Up plat to line (S)	UP,SFY	S
11a*	7&11	Line (S) to up plat	SFY,UP	S
11b*	7&10&11	Line (S) to down plat	SFY,DP	S
-	7	Release ground frame	GY,HS	Y

Table 5.1- *Note entries marked (*) apply ONLY when 7 is reversed.*

since the setting of points 12 will ensure that only one of them could be pulled anyway. However, with the shunt signals it is a different matter.

At the south end of the station, signal 11 (a, b or c depending on the setting of points 9 and 10) is clearly in conflict with signals 3 and 8, since these signal moves in opposite directions. At the other end of the station, matters are not quite so clear since we have used a single lever to fulfil several functions (4a, b). It would be permissible for 11 and 4 to be pulled together, for example to signal a shunting move right through the station, provided that *either* points 10 are normal and points 12 are reversed, *or* points 10 are reversed (and 9 normal) and points 12 are normal. However, to accommodate all such possibilities would lead to some extremely complicated conditional locking, so I decided that as a general rule, only one shunt signal could be 'off' at a time and that shunt signals should lock running signals and vice versa.

There is, however, one important exception to this general rule. We will often need to pull a train forward to clear points 6 when moving into or out of the goods yard. Therefore, we must allow signals 4 and 5 to be used together. However, this only applies to signal 4a (as selected by points 12), so we need a conditional rule here, that 6 locks 4 when 12 is normal.

We have mentioned the concept of sequential locking, and we will apply it here, by saying that 2 must be normal before 1 can be reversed, and similarly 14 must be normal

before 15 can be reversed.

Because we have used points to select which signal comes off when a given lever is pulled, it is important to prevent these points from being changed whilst a train is moving over them. To do this, we use both ways locks, so, for example, 3 and 11 should lock 9 and 10 both ways. Likewise, 4 should lock 12 both ways.

A similar situation applies to lever 7. Remember that we have used 7 (the ground frame release lever) to decide whether the yard operator or the fiddle yard operators drive when signals 3,4 and 11 are pulled off. If 7 were put back whilst one of these signals is off, then control would switch from one driver to another without warning, and the consequences could be disastrous. Therefore, 3, 4 and 11 need to lock 7 both ways. When 7 is reversed, the goods yard can be operated independently of the running lines. It therefore follows that the points allowing entry to the yard from the running lines, namely points 6 and 9, need to be locked by 7.

Applying these general principles, we might arrive at a list of locks and releases, as follows:

1   Locks 3,4,6,9,11,12
2   Locks 3,4,6,9,11 Releases (1W2N)
3   Locks 4,6,7BW,9,10BW,11
4   Locks 3,(6W12N),7BW,9,11,12BW
6   Locks 1,2,3,(4W12N),11,14,15 Releases 5
7   Locks 6,9

	1	2	3	4	5	6	7	8	9	10	11	12	13	14	15
1			L	L		L			L		L	L			
2			L	L		L			L	LR	L				
3	L	L		L		L	BW		L	BW	L	BW		L	L
4	L	L	L				BW		L	BW	L	BW		L	L
5															
6	L	L	L				L				L			L	L
7						L			L						
8											L				
9	L	L	L	L			L							L	
10		R													L
11	L	L	L	L		L	BW	L	BW	BW				L	L
12	L													R	
13															
14			L	L		L			L		L	LR			
15			L	L		L				L	L				

Table 5.2

8   Locks 11
9   Releases 8
10   Locks 15 Releases 2,9
11   Locks 6,7BW,8,9BW,10BW
12   Locks 1 Releases 14
14   Locks 3,4,6,9,11 Releases (15W14N)
15   Locks 3,4,6,10,11

How can we tell if our list is comprehensive, and that we haven't missed anything important? The answer is to use a little computer program that will be described in a later chapter. For the moment, let us proceed from the locking chart to the design of a set of mechanical locking bars that will do the required job.

From Chart to Locking Bars

Our goal is to produce a design for a set of locking bars that will meet all of the requirements of our locking chart, with as small a number of bars as possible. We do this by combining several locks on the same bar, and what follows is a methodology for doing this. Again, we shall look at a computer program later that will help to automate this process.

We exclude from the chart any adjacent lever releases, since these will be more easily done on the levers themselves. Thus 6, 10 and 9 releasing 5, 9 and 8 respectively do not need to be transferred to the chart. Because of the difficulty of combining conditional and sequential locks with conventional ones on the same bar, we exclude these from the chart also. We then transfer what is left in our list into a tabular form, as in Table 5.2.

When reading this table, we read from left to right along the rows, thus 1 locks 3, 4, 6, 9, 11 and 12. We also take the opportunity to ensure that all the mutual locks are completed. That is, when we fill in an 'L' in row 1 column 3, to mean that lever 1 locks lever 3, we also put an 'L' in row 3 column 1, to show lever 3 locks lever 1. A release means the releasing lever is locked in reverse, thus in row 10 column 2, we put 'R' to mean that 10 releases 2 and in row 2 column 10, we put LR to show that 2 locks 10 in reverse. Remember that both ways locks are not mutual. Thus, 3 locks 7 both ways, but 7 has no effect on 3. We have also added a couple of BW locks. Noticing that 3 needs to lock 10BW and 4 needs to lock 12BW, we can actually make both 3 and 4 lock 10BW and 12BW. This simplify things later as we shall see.

We now look for patterns. Basically we are trying to find rows where there are a lot of locks in common. These can all then be placed on the same bar. We immediately notice that rows 1, 2, 14 and 15 have a lot in common, namely that they all lock 3, 4, 6 and 11. 1, 2 and 14 also lock 9, but we notice that 15 locks 10. Since 10 releases 9, we can perfectly well have 15 lock 9 as well. So, let us make this our first locking bar: 1, 2, 14, 15 lock 3, 4, 6, 9, 11, as shown in figure 5.12.

We now update table 5.2 by either crossing out or ticking, or otherwise marking those locks which we have now accommodated on bar A. Remember that we tick off both the lock and its counterpart, i.e. row 1 column 3, and row 3 column 1 etc. Our updated table now looks something like Table 5.3.

In this case, the locks that have now been satisfied have been converted to italics. Next, we observe that rows 3 and 4 have much in common, in particular a lot of 'both ways'

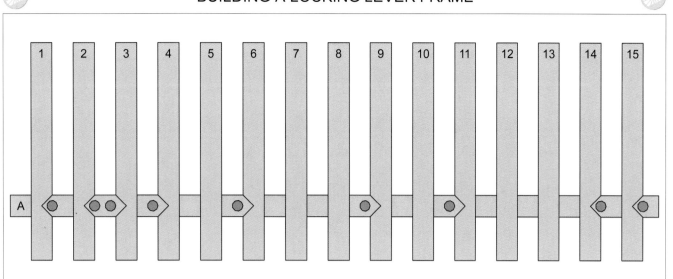

Above Figure 5.12.   *Below Table 5.3*

	1	2	3	4	5	6	7	8	9	10	11	12	13	14	15
1			*L*	*L*		*L*			*L*		*L*	**L**			
2			*L*	*L*		*L*			*L*	**LR**	*L*				
3	*L*	*L*		**L**		**L**	**BW**		**L**	**BW**	**L**	**BW**		*L*	*L*
4	*L*	*L*	*L*				**BW**		**L**	**BW**	**L**	**BW**		*L*	*L*
5															
6	*L*	*L*	**L**				**L**				**L**			*L*	*L*
7						**L**			**L**						
8											**L**				
9	*L*	*L*	**L**	**L**			**L**							*L*	*L*
10		**R**													**L**
11	*L*	*L*	**L**	**L**		**L**	**BW**	**L**	**BW**	**BW**				*L*	*L*
12	**L**													**R**	
13															
14			*L*	*L*		*L*			*L*		*L*	**LR**			
15			*L*	*L*		*L*			*L*	**L**	*L*				

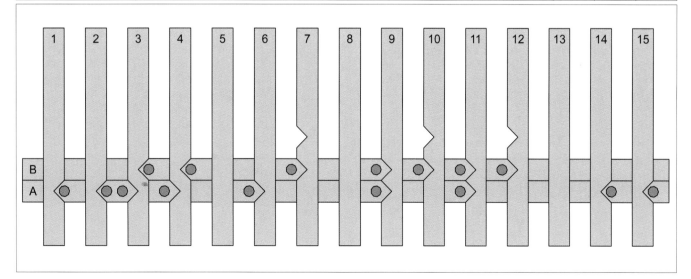

Figure 5.13

	1	2	3	4	5	6	7	8	9	10	11	12	13	14	15
1			*L*	*L*		*L*			*L*		*L*	**L**			
2			*L*	*L*		*L*			*L*	**LR**	*L*				
3	*L*	*L*		**L**		**L**	*BW*		*L*	*BW*	*L*	*BW*		*L*	*L*
4	*L*	*L*	**L**				*BW*		*L*	*BW*	*L*	*BW*		*L*	*L*
5															
6	*L*	*L*	**L**				**L**				**L**			*L*	*L*
7						**L**			**L**						
8											**L**				
9	*L*	*L*	*L*	*L*			**L**							*L*	*L*
10		**R**													**L**
11	*L*	*L*	*L*	*L*		**L**	**BW**	**L**	**BW**	**BW**				*L*	*L*
12	**L**													**R**	
13															
14			*L*	*L*		*L*			*L*		*L*	**LR**			
15			*L*	*L*		*L*			*L*	**L**	*L*				

Table 5.4

locks, so let us make our next locking bar (bar B) this: 3, 4 lock 7BW, 9, 10BW, 11, 12BW (figure 5.13).

Notice that the tappets that are to be locked both ways have two notches – one in line with bar B when the lever is normal, and one which will line up when the lever is reversed. It is *most important* that the travel of the tappets from the normal to the reverse position is not a whole number multiple of the locking bar width. If it were, this could bring notches into line with locking dogs and prevent a lock from occurring. Look at tappet 9, for instance. If the travel were exactly the same as the locking bar width, then

reversing 9 would not lock 1, 2, 14 or 15, since the upper notch would now line up with the lock in bar A. In our case, we have made the travel 1½ times the locking bar width, but it is well to keep an eye on these things. Updating our table again, we now have Table 5.4.

Notice how quickly the number of unsatisfied locks goes down. With just two locking bars, we have reduced the number to well under half. We shall not test the reader's patience with all the individual steps that remain, but cut immediately to the conclusion. The table is fully satisfied by seven locking bars, as follows:

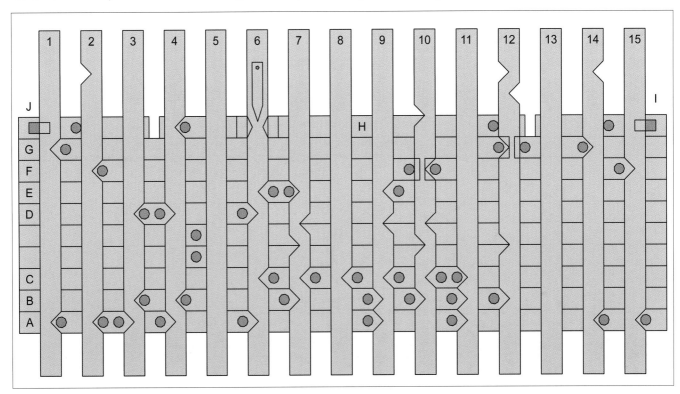

Figure 5.14

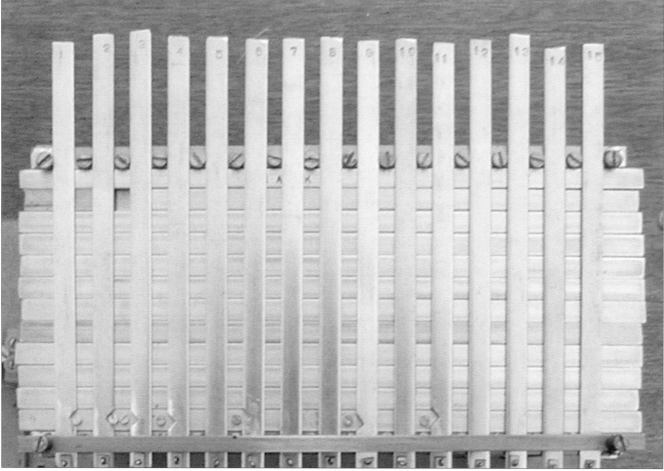

Figure 5.15 *– Locking bar A in construction*

A 1,2,14,15 Lock 3,4,6,9,11
B 3,4 Lock 7BW,9,10BW,11,12BW
C 11 Locks 6,7BW,8,9BW,10BW
D 3 Locks 4,6
E 7 Locks 6,9
F 10 Locks 15 Releases 2
G 12 Locks 1 Releases 14

In addition to these, we need two bars (I and J) for the sequential locks on levers 1 and 15, plus a further bar (H) for the conditional lock, 6 locks 4 when 12 is normal. Now, you may wonder how this conditional lock can be done. The condition is on lever 12, but if we place the swinging link on tappet 12, how can it be intermediate between 4 and 6? The answer is, we slightly rephrase the condition. If you think about it, saying that 6 locks 4 when 12 is normal is exactly the same as saying that 12 releases 4 when 6 is reversed. Implementing the latter is straightforward: we place the swinging link on tappet 6, and have the usual locks and notches for a release on tappets 12 and 4. Figure 5.14 illustrates the complete set of locking bars.

Notice that we have been able to fit bars H, I and J alongside each other, since I and J only require the very ends of the locking tray. Also notice that two dummy locking bars have been inserted between bars C and D. This is in order to prevent notches from interfering with each other. You may find, depending on particular circumstances, that juggling around the order of the bars can achieve the same result. Bars F and G are split in two. This is so that when 10 is pulled, the right hand half of bar F moves right, locking 15, whilst the left hand half remains where it is until 2 is pulled, locking 10 in reverse. Similar considerations apply to bar G.

Constructional Notes

Having designed our bars, we now need to build them. The key here is to take things a step at a time. Do not mark out and file all the tappets, then drill and tap all the locking bars – they will never match up! Make one locking bar, file its corresponding notches, then test it thoroughly before making the next.

The locking tray itself can be any flat surface that will stay flat over a period of years. Do not use plywood or similar, this will warp. I use $\frac{1}{16}$" brass sheet, although reasonably thick MDF (9-12mm) will do also. For the locking bars, rectangular brass bar is suitable; I use $\frac{1}{4}$" x $\frac{1}{8}$" for these. This thickness gives you enough metal to make a reasonable threaded hole for the locks. For these, I use 6BA or 8BA set screws. The locking bars should slide easily across the surface of the locking tray. To ensure this, fix a length of $\frac{1}{4}$" x $\frac{1}{8}$" brass to the locking tray as a guide. Next place the required number of lengths of bar against it, and finally fix a second length of bar to the locking tray, such that all the bars slide freely, but without too much side play.

For the tappets, slightly thinner material will do. I use $\frac{1}{4}$" x $\frac{1}{16}$" brass flat for these. Conveniently, this width of bar plus the diameter of the head of a 6BA set screw comes out almost exactly the same as the lever spacing on a twelfth scale model of a Stevens lever frame! You can fit the 6BA screws into threaded holes on the two fixed guides in order to space the tappets accurately (see figure 5.15). You will

need two further brass bars above the tappets to keep them in place. In figure 5.15 only the lower one of these is fitted; in figure 5.16, both are in place.

Start with one locking bar, and mark out and file the first 'notch' in a tappet. Then, mark where the centre of the hole for the lock is to be, on the centre line of the locking bar. Drill and tap this for the set screw, fit the latter, then file off any excess from the back of the locking bar. Fit all back together and check that moving the tappet causes the locking bar to slide smoothly in the required direction. In figure 5.15 the first few locks on bar A (1, 2 lock 3, 4, 6, 9, 11) have been completed.

Proceed in this way with each of the locks. Make sure that you test each one as you make it. Common problems are notches not filed deep enough, holes for lock screws not quite on the centre line of the locking bar, or the sides of the notches not smooth and straight enough to push the locks sideways. Fix each one as you make it, and do not proceed to the next until you are sure it works.

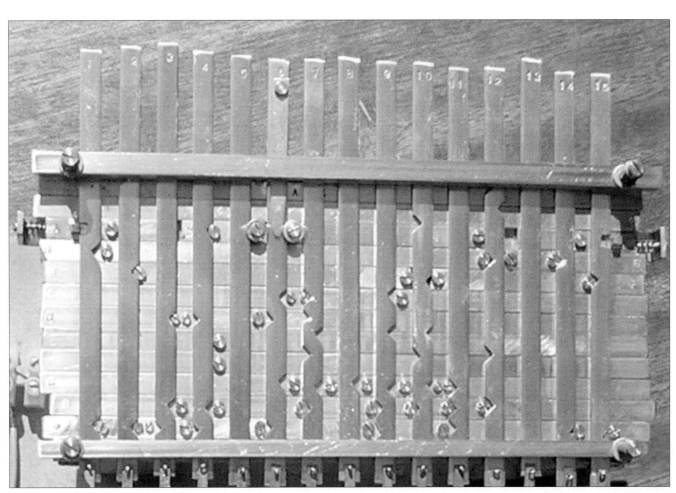

Figure 5.16 – *The finished locking tray*

Figure 5.17 *– Sequential locking*

For 'release' notches and those associated with 'both ways' locks, you will have to position the notch in the tappet such that it lines up with the lock when the associated lever is in the reverse position. Measure this carefully from your lever frame – it is useful to scribe a line on each of the tappets to show its position (relative to the fixed guides) when the lever is normal and reversed. Remember it is important not to have the travel distance a whole number multiple of the width of the locking bars.

Once the locking tray is completed, attach it to the lever frame. A suitable linkage is a short length of 3mm x 1mm brass twisted through 90°. Fix it to the lever and tappet with split pins. Use the thinnest split pins you can find. They should be opened out only slightly, such that the link between lever and tappet can be easily broken. The reason for this is that they are much easier to replace than a broken lever or a bent tappet. You will invariably find an operator who is determined to pull a particular lever by brute force, even when the locking is trying to prevent it! The split pin is what engineers call a sacrificial component, it is there to prevent damage to other components which are much more expensive or difficult to replace. Figure 5.16 gives a view

of the completed locking tray. Notice that where two locking dogs are required between tappets the space is very tight. For example, on locking bar A between tappets 2 and 3, and again on locking bar C between tappets 10 and 11, we need a pair of locking dogs. To fit these in, we use 8BA set screws instead of 6BA.

Figure 5.17 illustrates the sequential locking in operation. Notice that lever 2 has been pulled, thus locking 1 in the normal position until such time as 2 is restored to 'danger'. Notice also in this picture that because lever 6 is normal, lever 4 is free to move independently of 12. Were 6 to be reversed, the pivoted arm would come forward, thus transferring the movement of the locking bar associated with 4 to that of 12, thereby providing the conditional locking required.

A view of the completed lever frame, mounted in its box, can be seen in the colour plates section.

6

RELAY LOCKING

The transition from mechanical operation of points and signals to power assisted operation meant that the signalman no longer had to rely on his own muscle power to move point blades or signal arms. The need for large levers to give sufficient mechanical advantage thus disappeared, and miniature levers were used in the early power frames. Initially, these were locked mechanically, just like the full-size levers of a standard lever frame. However, it was soon realized that if all the lever needed to do was to close a pair of electrical contacts, then an ordinary electrical switch would do just as well, and this could be locked electrically by relays instead of by a mechanical system of tappets and locking bars. Thus in the 1920s began the trend toward relay locking.

Unlike mechanical locking, where the movement of a lever is physically obstructed by the locking mechanism, relay locking works by depriving the miniature lever or switch of electrical power. However, the operation of the switch was not physically prevented. Therefore, some means of indicating to the signalman whether or not a given switch was operational or not was needed. A standard method was to provide a light above the switch which was illuminated if the switch was 'live', that is, if the signal it operated was free to move.

In the early power frames, rows of switches replaced the row of levers. The switches, however, still had a corresponding function to the levers they replaced – each would operate a point or signal, and would also affect the ability of other switches in the frame to operate. Switches would be colour-coded in a similar fashion to the levers, red for signals, black for points etc. To all intents and purposes, therefore, the row of switches corresponded exactly with the row of levers in a mechanical frame.

It may seem at first sight that the whole idea of using switches instead of levers runs counter to the mantra of this book: *no switches*. However, it may be thought acceptable to use switches where the prototype did so, as in a miniature power frame. Perhaps our mantra should be *no extraneous switches*!

Even if you decide to use a mechanical frame with tappet locking, there is one situation in which relay interlocking might be used alongside. This is where a lever on one frame is to lock or release a lever on another remote frame. You can achieve this mechanically, for example by means of a solenoid moving a locking bar or tappet. However, it is far easier to do this electrically, simply by depriving the appro-

priate lever of its electrical power by means of a relay.

In what follows, we will speak of a switch being 'open' when it is in the normal or 'off' position, and 'closed' when it is in the reverse or 'on' position. In diagrams, all switches and relays are shown in their normal or 'off' positions. The contact which is connected to common in this position is called 'normally closed' (NC), that which is not so connected is called 'normally open' (NO). We now look at how the various styles of lock and release can be achieved using relays instead of mechanical tappets and locking bars.

A Locks B

Let us start with the simplest case. Switch A is to lock switch B. Obviously, this is to be mutual: switch B is to lock switch A also. To accomplish this, we simply ensure that the electrical feed to each of the two switches goes through a 'normally closed' contact of the alternate switch's relay. To illustrate, in figure 6.1 we have two switches, SW1 and SW2, which lock one another.

When neither relay is closed, +12v is fed to both switches via the 'normally closed' contact of the alternate relay. Both switches are therefore available, and this is indicated by their corresponding LED being illuminated. Note that the LEDs shown are of the 12v variety with a built-in current limiting resistor. If you use standard LEDs you will need a 1K resistor in series with them. Now, suppose we close one of the switches, say SW1. Immediately, relay RL1 closes and +12v is no longer available to switch SW2. The LED associated with SW2 goes out, indicating that this switch is no longer 'live'. Even if it were to be closed, relay RL2 would not operate. In the same way, if SW2 were to be closed first, it would be impossible to operate relay RL1. To drive further circuitry, e.g. logic relays or track relays we would use outputs taken from additional poles on relays RL1 and RL2.

A Releases B

Suppose we now wish to have switch SW1 release SW2. That is, SW2 is to be non-operational until SW1 is closed. With mechanical levers, the releasing lever becomes locked in reverse once the released lever is pulled. Correspondingly we wish to ensure that if SW2 is closed (having first been released by SW1), it is then impossible to put SW1 back to normal whilst SW2 remains reversed. Of course, we do not mean that the switch itself cannot be turned off, but that the state of the relays cannot be changed by restoring

Figure 6.1 - *SW1 locks SW2*

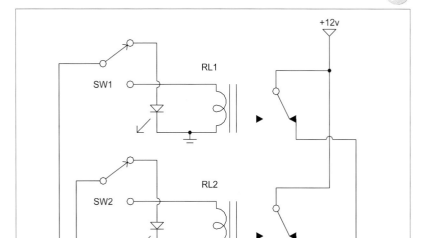

Figure 6.2 - *SW1 releases SW2*

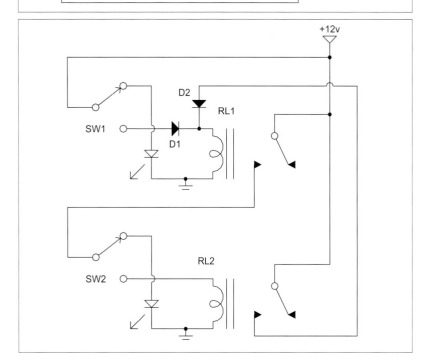

SW1 to normal. Remember that our 'output' from these circuits is always to be taken from the relays via an additional pole.

In figure 6.2, we see that in the position shown with both switches 'off', the LED associated with SW2 has no electrical feed. It is thus off, indicating that SW2 is non-operational: if SW2 were closed, nothing would happen. However, if SW1 is closed, current is fed via diode D1 to the coil of relay RL1 which closes, feeding +12v to SW2 and its LED. SW2 is now live, and if it is closed relay RL2 will operate. As it does so, +12v is fed via a second diode, D2, to the operating coil of RL1. The purpose of these diodes is to provide an 'either-or' function. If *either* SW1 is

closed, *or* RL2 is closed, then RL1 will operate. This means that even if SW1 is now returned to the 'off' position, RL1 will stay closed until such time as SW2 is restored to normal also. Thus SW2 has the effect of locking SW1 in reverse, as required.

A Locks B Both Ways

Here, we wish to ensure that when A is reversed, B stays in whatever state it is currently in, i.e. if B is 'off' at the time A is reversed, then B cannot be put 'on', and if B is 'on' at the time A is reversed, then it cannot be turned 'off'. Both restrictions continue for as long as A remains 'on'.

'TRAX 3' SIGNALLING and LEVER FRAMES

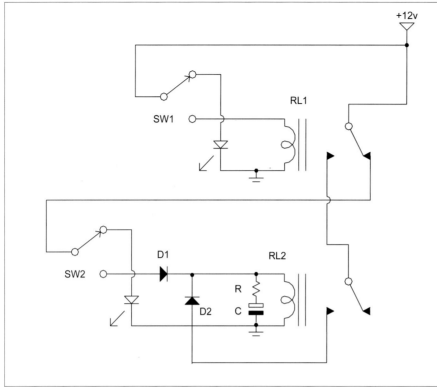

Figure 6.3 *- SW1 locks SW2 both ways*

In figure 6.3, we see that, so long as SW1 remains 'off', SW2 is supplied with +12v, and can thus turn its relay, RL2, on and off at will. Suppose, for the moment, that SW2 is 'off' (as shown). If we now turn SW1 'on', then relay RL1 will close. SW2 now has no +12v supply, and thus cannot operate relay RL2. Hence, RL2 remains 'off' for as long as SW1 is 'on' – it is locked in the 'off' position.

Now suppose that SW2 is 'on' just before SW1 is turned 'on'. Relay RL2 will thus be closed, being supplied with +12v via RL1's normally closed contact, SW2 and the diode D1. When RL1 closes, Relay RL2 will still be supplied with +12v via the second diode D2, its own normally open contact and that of RL1. Thus RL2 will stay closed, even if SW2 is restored normal. In other words, it is locked in the 'on' position.

However, the astute reader will notice that, during the very small period of time in which the contacts of RL1 change over, both sources of supply of +12v to relay RL2 will be cut off. The purpose of resistor R and capacitor C is to prevent RL2 opening during this time. C stores sufficient charge to hold RL2 closed for long enough to bridge the gap, so to speak. The actual values of R and C will depend on the holding current of the relay and its release time. You can find these by experiment. 100 ohms and 100 micro-farads is a reasonable starting point.

Conditional Lock

As we saw in the previous chapter, there are occasions when conditional locking is required, i.e. A locks B, exactly as in figure 6.1, but only when C is reversed.

In figure 6.4 we have shown a situation where SW1 and SW2 are independent whilst SW3 is open. Both receive +12v via RL3's normally closed contacts. However, if SW3 is closed, thereby closing RL3, it will be seen that +12v is now fed to each of switches SW1 and SW2 via the normally closed contacts of each other's relays, as in figure 6.1. Thus SW1 locks SW2 (and vice versa) when SW3 is reversed.

Fiddle Yard Example

The following simple example will illustrate how the above general principles of relay locking can be put into practice. The example is in fact a fiddle yard for 'Abbotstone', a club layout built by the members of Winchester Railway Modellers. Trains are assembled on the storage roads, and make their appearance onto the scenic part of the layout through a tunnel mouth. When trains return to the fiddle yard, they are placed in one of the storage roads, the engine uncoupled and run onto a cassette, which is then turned, and offered up to an empty storage road for the engine to run back onto its train.

Referring to figure 6.5, points 2, 3 and 4 give access to and from the storage roads for trains passing through the tunnel. However, we do not wish to use points 2 and 3 when running an engine onto its train. This might result in the engine sticking its nose out of the tunnel onto the scenic part of the layout, then popping back into the tunnel, which would look a little odd, to say the least! Therefore, points 5 and 6 are provided so that engines can shuffle back and forth from road to spur and spur to road without any danger of appearing in the tunnel mouth. This does mean that storage road B must only be used for short trains. Points 7 give access to some other storage facilities (basically a sector plate).

We wish to provide suitable locking for the above to prevent conflicting switch settings. The locks required are as follows:

1 Locks 2BW,3BW,4BW,5,6,7,8
2 Locks 3,6
3 Locks 2,5

62

4 Locks (1W2N),5,6,7,(8W2N)
5 Locks 1,3,4,6,7,8
6 Locks 1,2,4,5,7,8
7 Locks 1,4,5,6,8
8 Locks 1,2BW,3BW,4BW,5,6,7

The incoming and outgoing signals, 1 and 8, lock points 2, 3 and 4 both ways to prevent points being changed under a moving train. They also lock points 5, 6 and 7 normal. Points 5 and 6 are used to shuffle an engine between spurs E and F and road B. Similarly, points 4 are used on their own (i.e. not in conjunction with points 2) to move an engine between spur F and road D. Suppose we have signalled an incoming train into road A, with switches 3 then 1. We uncouple its engine, and turn it on the cassette. Suppose road D is empty. We would then put the engine on road D, and using points 4, move it onto spur F. Then, restoring 4, we would reverse 6 to move our engine onto road B. Next, restore 6 and reverse 5 to move the engine onto spur E, and finally, restore 5 to normal to back the engine onto its train.

Switch Indicators

Above the switches on figure 6.5 we have shown circles to represent LEDs. As has been mentioned, there is no way of preventing a switch from being changed, however we can prevent it from being effective by removing its

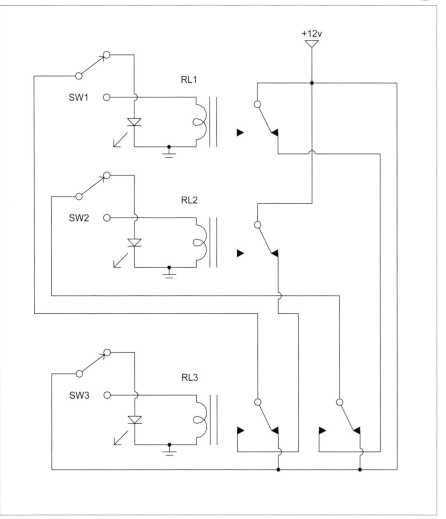

Above - Figure 6.4 - *SW1 locks SW2 when SW3 is reversed*

Below - Figure 6.5 - *Fiddle Yard Example*

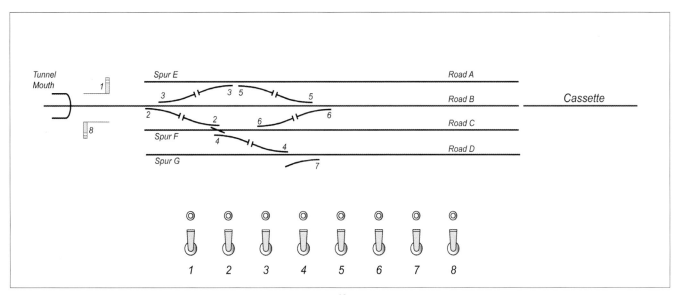

power source. The LEDs indicate which switches are actually operational. If an LED is on, then its corresponding switch is available. If an LED is off, the switch can be clicked, but will have no effect.

Locking Details

As we have seen, the general principle of relay locking is that +12v is fed to the 'common' side of each switch via a set of relay contacts, the relays in question being operated by the other switches involved in the locking rule. For example, switch 1 is to be locked by 5, 6, 7 and 8. Its +12v supply is therefore fed via the 'normally closed' contacts of relays RL5, RL6, RL7 and RL8, as shown in figure 6.6 (note that, as per convention all relay and switch contacts are shown in the 'normal' position).

To simplify understanding of the logic, the normal practice of drawing the relay contacts alongside the relay coil (as in figures 6.1 – 6.4) has not been followed. Instead, the contacts are drawn in line with the switch they are involved in locking. Below each set of contacts, the relay number to which they belong is indicated, it being understood that the contacts are in the normal condition, i.e. with the relay coil un-energised. To distinguish relay contacts from switches, the former are drawn as with pointed NO and NC contacts, whilst switches are drawn with round contacts.

In figure 6.6, it will be seen that +12v is fed to the common side of switch 1 if, and only if, all of the relays 5-8 are in the 'normal' position. As soon as any link in this chain is broken, switch 1 no longer has +12v and its LED goes out.

A very similar chain of logic applies to switch 8. This is to be locked by 5, 6, 7 and 1. However, noting that 5, 6, and 7 are common to both switches 1 and 8, we can save some relay contacts by sharing 5, 6 and 7 with both, as in figure 6.7. We also add the conditional locks on switches 1 and 8. 4 is to lock both of these when 2 is normal (when 2 is reverse, this is a perfectly valid combination). To accomplish

this, we note that in figure 6.6 +12v is permanently connected to the 'chain' of relay contacts on the left of the diagram. If, instead of this permanent connection, we supply +12v via two further relay contacts, operated by switches 2 and 4, we can make this conditional lock, as shown in figure 6.7. When 2 is reversed, +12v is supplied via its NO terminal. When 2 is normal, 4 must be normal also to feed +12v to the 'chain' of contacts. Thus, 4 locks 1 and 8 when 2 is normal, as required.

We see here why it is important to distinguish between relay contacts and switches. In locking switch 1, it is the relay contacts on the relay associated with switch 8 that are used. In locking switch 8, it is the relay contacts on the relay associated with switch 1 that are used. This ensures that, if switch 1 is reversed, then the chain feeding +12v to switch 8 is broken. Relay 8 cannot therefore be activated, even if switch 8 is reversed, and the 'normally closed' relay contact in the chain feeding switch 1 cannot be broken. In other words, once switch 1 is reversed, switch 8 is non-operational and this will be indicated by its LED going out.

We turn now to the 'both ways' locking required for switches 2, 3 and 4. Whilst switches 1 and 8 are normal, 2, 3 and 4 can be changed at will. However, when either 1 or 8 is reversed, 2, 3 and 4 will be locked in whatever position they happen to be until such time as both 1 and 8 are set normal.

In figure 6.8, we show the both ways locking for SW2. Notice that we have combined the signals 1 and 8 together using a couple of diodes and relay RL9. If neither 1 nor 8 is reversed (i.e. both sets of contacts 1 and 8 are as shown), relay RL9 will be off, and +12v is present at point 'A', so switch SW2 will operate. However if either 1 or 8 is closed, one or other of the two diodes will conduct, causing relay RL9 to close, removing the +12v supply from point 'A'. However, +12v will now be available at point 'B', so if relay RL2 is closed at this point, it will remain closed. If relay RL2 is open, it will remain so, and cannot be closed by

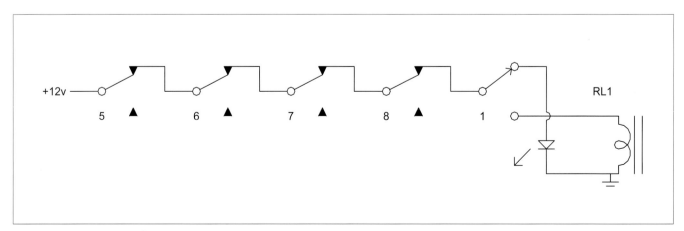

Figure 6.6 - 5,6,7,8 Lock 1

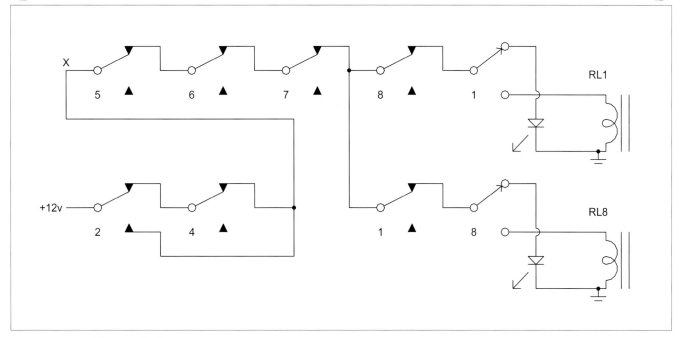

Figure 6.7 - *Adding Locks for 8*

means of SW2 until both 1 and 8 are returned to normal. Thus 1 and 8 lock 2 both ways, as required. As before, the capacitor ensures that relay RL2 stays closed during the few milliseconds it takes for relay RL9 to change over, there being no volts at either 'A' or 'B' during this brief time.

This is not the whole story, however, since we want 2 to be locked by 3 and 6 also. This is easily accomplished by adding further relay contacts (on relays RL3 and RL6) between point 'A' and switch SW2 (see figure 6.9). Switches SW3 and SW4 are treated likewise. Each are to be locked both ways by 1 and 8, and in addition, 3 is to be locked by 2 and 5, and 4 is to be locked by 5, 6 and 7. Again, these locks are accomplished by placing relay contacts in series with the supply of +12v to their respective switches. Note, however, that this source of +12v is taken from the 'normally closed' (NC) contact of relay RL9, corresponding to point 'A' of figure 6.8, so that these switches only operate when neither 1 nor 8 is reversed. Also in figure 6.9, we have added the lockings required for 5, 6 and 7. 5 is to be locked by 3, 4, 6 and 7. Noting that we already have 3 and 6 locking 2, we can use these relay contacts in the 'chain' for locking 5. Similar considerations apply to switch numbers 4 and 6.

Figure 6.9 thus represents a complete circuit for locking switches 1 to 8 by means of relays. The reader might like to trace through this circuit, checking each of the eight locking rules given above to see how each rule is implemented.

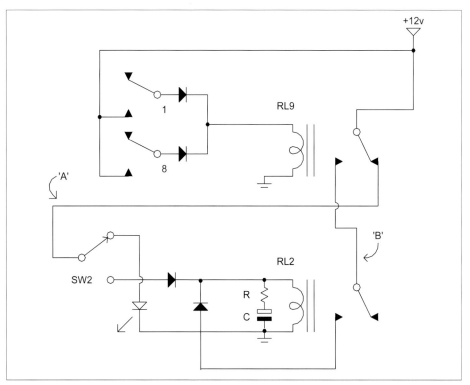

Figure 6.8 - *1 and 8 lock 2 both ways*

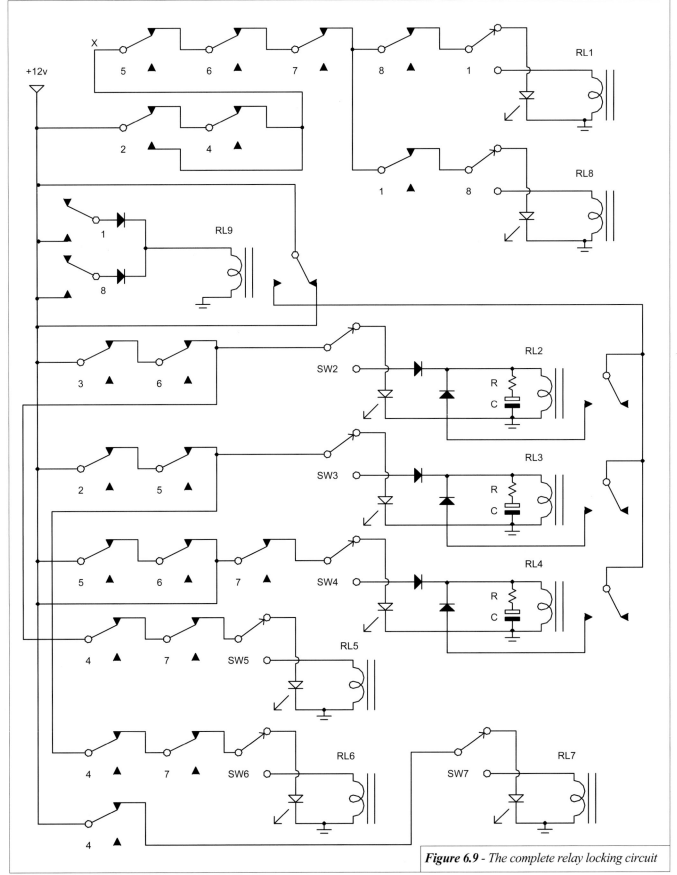

Figure 6.9 - The complete relay locking circuit

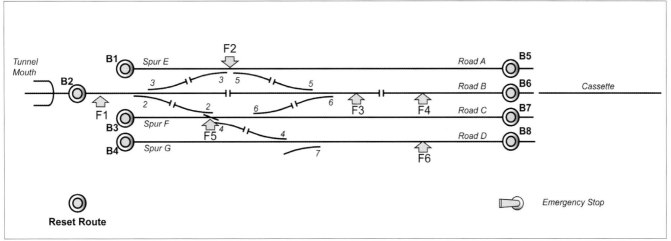

Figure 6.10 - *An 'NX' style panel*

An 'NX' Panel

The fiddle yard described above was eventually rebuilt in a form resembling that of an Entrance-Exit panel, as described in chapter 2. Although not a full 'NX' implementation, the panel incorporates the general principles of NX control, in that the user presses just two buttons to indicate where the train is to move FROM and where it is to move TO. Relays in the panel then set the points appropriate for this route, connect the necessary track feeds to the train controller and allow the operator to make the move.

With reference to figure 6.10, routes are set by push-button switches, B1 to B8, operated in pairs. A route is set by first pushing the button corresponding to where the train is to move FROM, then pushing the button corresponding to where it is to move TO. For example to move a train from Abbotstone into road A, we would push B2 then B5. A route is cancelled and the system set up for a new route by pushing the RESET button.

When button B2 is pushed as the second one of a pair, control is handed to the 'Abbotstone' controller after the route has been set. In all other circumstances the Fiddle Yard controller has control. LEDs (shown in figure 6.11) adjacent to each button illuminate when the button is pressed. When the second button of a pair is pressed, LEDs on each crossover to be reversed and on each feed connected are also illuminated. These are bicolour LEDs – if the fiddle yard controller is in charge, they are green; if Abbotstone, red. The 'emergency stop' switch disconnects power from Abbotstone and is used if there is a derailment or other problem of which the Abbotstone operator could well be unaware. Note that this leaves the route set (unlike the reset button) so that when the problem is fixed it can simply be returned to the normal position.

For reference to the circuit diagram, figure 6.11 numbers the LEDs. Numbers 1-8 correspond to buttons 1-8, and come on immediately the button is pushed. LEDs 9-14 correspond to track feeds F1 to F6 and will only come on if a valid pair of buttons has been pushed. LEDs 15 to 19 correspond to point motors PM2 to PM6 and will come on as required to set a

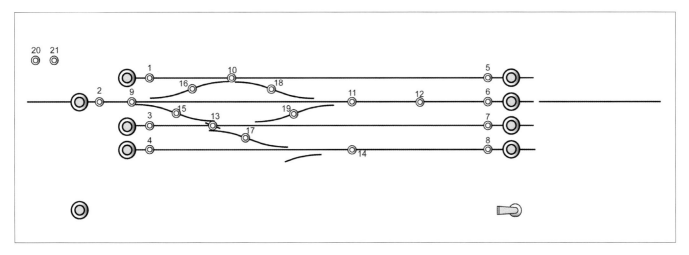

Figure 6.11 - *LED Indicators 1 to 21*

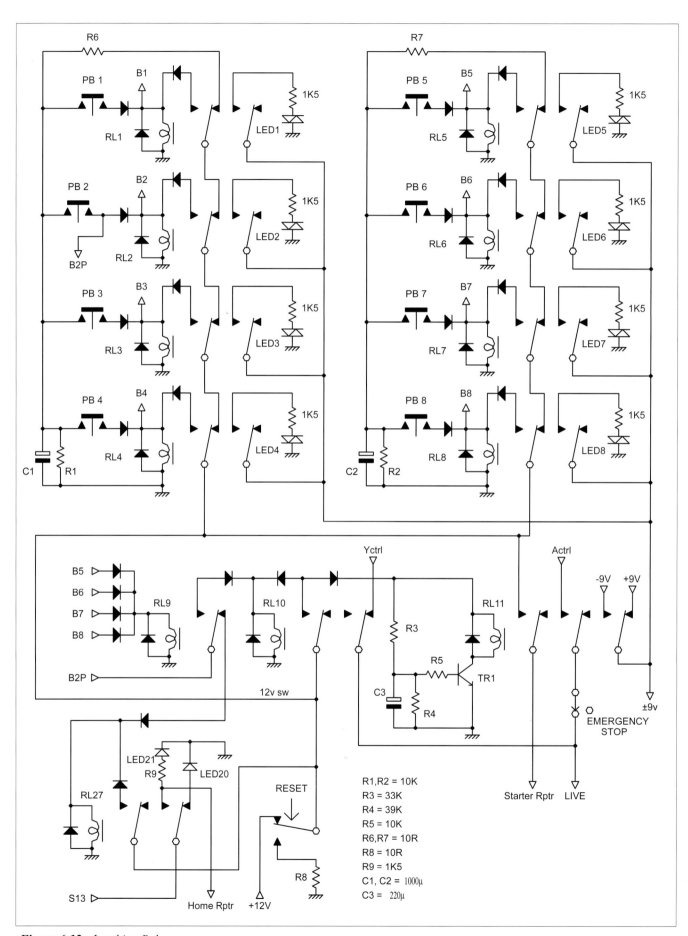

Figure 6.12 - Latching Relays

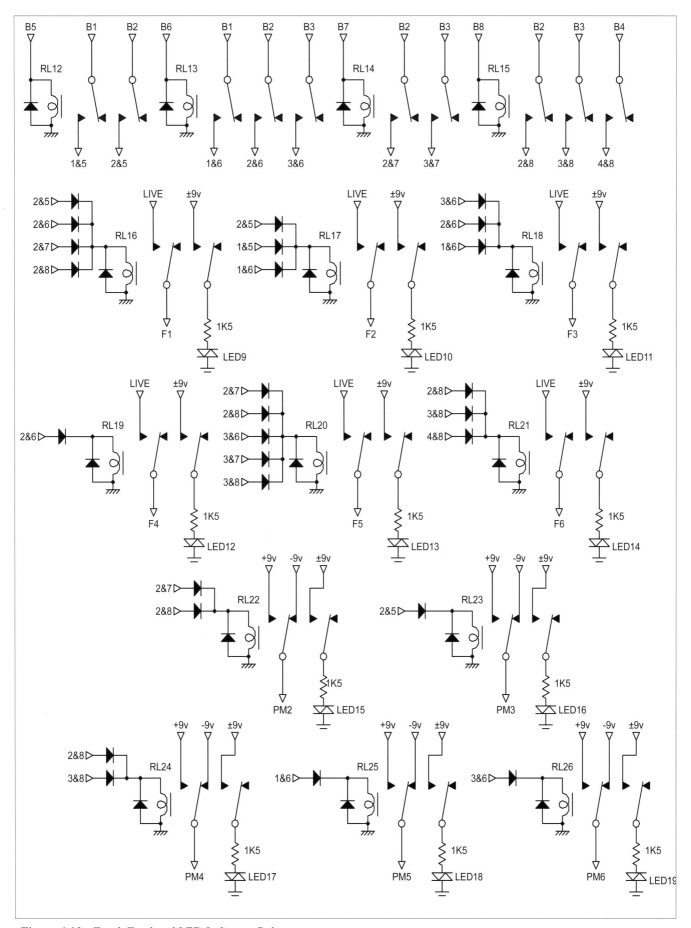

Figure 6.13 - *Track Feed and LED Indicator Relays*

route between two valid buttons. Finally, LEDs 20 and 21 are those operated by relay RL27 to indicate the readiness of an incoming train from Abbotstone.

The Circuit Diagram

Figures 6.12 and 6.13 give the circuit diagram for the panel. Figure 6.12 contains the logic for deciding which buttons have been pushed in what sequence. Buttons PB1 to PB4 are push-to-make switches. Each turns on a relay (RL1 to RL4) which then latches 'on', preventing any other button in the group from working by removing the +12v feed to the buttons. The relays provide latched outputs B1 to B4. Capacitor C1 holds enough charge to ensure that the selected relay latches securely after the +12v feed is removed. Button PB2 also has a non-latched output, B2P (for B2 pulsed) whose purpose will be revealed shortly.

Push buttons PB5 to PB8 form a corresponding group with a similar function.

The circuitry around RL9-RL11 determines which controller is to be used. If one of the group of buttons B5-B8 is pressed first, then B2 is pressed second, the 'momentary' signal B2P causes relay RL10 to latch 'on'. This, in turn, disconnects the fiddle yard controller (Yctrl) from the 'LIVE' feed and, through R3, causes C3 to begin to charge up. Once the voltage at C3 reaches 0.6v, TR1 starts to conduct, turning on relay RL11. This connects the 'Abbotstone' controller (Actrl), and also reverses the polarity of the feed to the LEDs. The delay is to ensure that point motors have fully changed before power is connected to Abbotstone.

The RESET button disconnects +12v from all relays, causing them all to come off. The EMERGENCY STOP switch temporarily disconnects Actrl without changing any points etc.

Figure 6.13 shows how the B1 – B8 outputs of figure 6.12 (from relays RL1 to RL8) are used to with relays RL12-RL15 carry out the logic required to specify each valid route setting (1&5, 2&5, etc). These are then used to switch relays RL16-RL21 to control track feeds, and relays RL22-RL26 to set points, along with their corresponding LEDs.

Figure 6.14 is a view of the completed 'NX' style panel. Notice that the 'Reset Route' push-button and the 'Emergency Stop' switch (cf figure 6.10) are labelled 'O' and 'X' respectively. This is not to be obscure, but because the layout is sometimes operated from the front (i.e. in the clubroom) and sometimes from behind (at a show). The beauty of 'O' and 'X' is that they look the same upside down. All we have to do is hang the box on the appropriate side of the layout and the panel can be read either way up. The fact that the push-buttons are used solely to indicate locations on the track plan (i.e. move from *here* to *there*) means that they do not need any labelling. However, we did decide that there should be some method of referring to particular buttons. Ideally, we would have used differently coloured round push-buttons. However, the club spares box happened to have eight square red buttons so we used these. The oval panels alongside the buttons are in fact colour-codes so that we can distinguish between them.

Notice that, by recognizing only those button combinations that are valid, we have performed the same function as mechanically locking invalid lever combinations in a mechanical lever frame. For example, referring to figure 6.10, pressing B4 then B5 would require a route from the bottom left spur to the top right storage road. This cannot be accomplished in a single move, so we simply do not recognize it – you will notice that at the top of figure 6.13 there is simply no relay detecting the combination 4&8. Thus the user can push any sequence of buttons he likes, but the circuit will only respond to valid combinations.

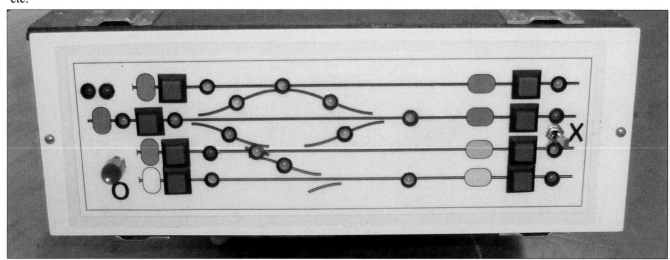

Figure 6.14 The 'NX' style panel

7

INTRODUCING *TRAX*

Accompanying this book is a CD containing the third version of the *Trax* program. This is a program designed to run on a PC under the Windows operating system. It provides facilities to help you design a layout, plan and test your electrical connections, print templates to allow you to build your own track, design interlocked lever frames and many other tasks.

If you are already familiar with *Trax 1* or *Trax 2*, you might wish to proceed directly on to the next chapter, which concentrates on those features which are new in *Trax 3*. However, if you are not one of the several thousand people who have an earlier version, you might find it useful to first have a look through this chapter, which will take you through the basics of developing and testing a layout design.

Getting Started

Place the disc in your CD reader, and run the command "D:setup.exe" where D is the drive letter of your CD reader. You will be asked to agree to the terms and conditions of the software license and *Trax 3* will then be installed.

Start up *Trax 3* You will notice that most of the menu options and buttons are grayed out. This is because we do not yet have a layout to work on. Click File | New and the 'New Layout' dialog box will appear (figure 7.1). Enter a size of 18 feet by 3 feet. Click OK, and you should have a nice long layout with 1 ft gridlines.

The *Trax* Screen

Now that we have a layout, most of the elements of the *Trax* screen become active (figure 7.2). At the top of the screen are the standard caption and menu bars. Below this, to the left, are a number of Toolbuttons, which provide quick access to commonly required menu items. Hover your mouse cursor over these buttons to see what they do. To the right

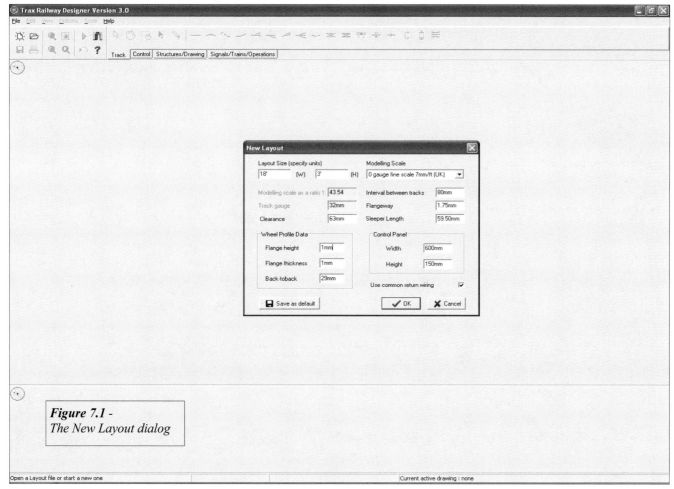

Figure 7.1 -
The New Layout dialog

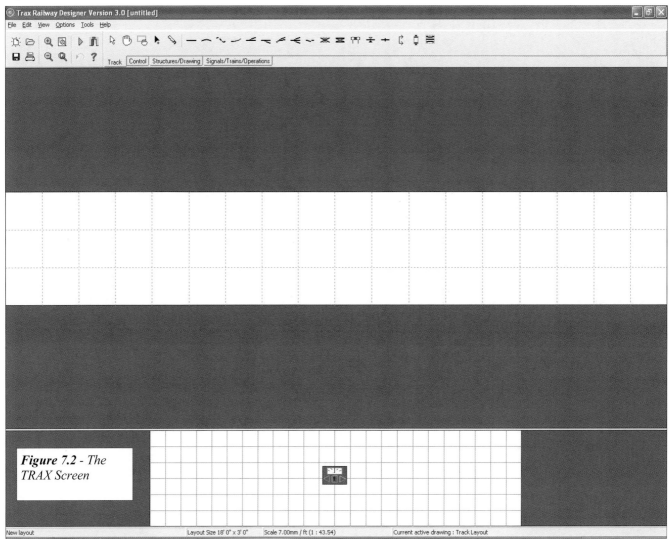

Figure 7.2 - The TRAX Screen

are a series of Palettes. Each Palette provides a number of buttons which enable you to add particular parts to your layout. On the Track Palette are straights, curves, points, crossings, feeds etc. On the Controls Palette are the lever frames, switches, buttons and train controllers. On the Structures /Drawing Palette are various tools to enable you to draw buildings, platforms etc on the layout. Finally, on the Signals/Trains/Operations palette are semaphore and colour light signals, also the tools to assist in developing and testing a timetable.

The main window area is divided into two parts. In the upper part is the layout track plan, and in the lower part is the control panel. You can drag the dividing line between these two up and down so that, for example, when you are working on the track plan, you can minimise the space occupied by the control panel. Finally, along the bottom of the screen is a status bar, which will prompt you from time to time with a message or line of information.

Adding and Joining Track Parts

Click the 'Straight Track' button on the 'Track' palette and then click the mouse to put three lengths of straight track roughly in the central part of the layout. Next, join them up. On the Track palette, click the Joining tool. Then click one end of the middle straight. It should be highlighted in purple, and the end to be joined indicated by a small cross. If it isn't, you didn't quite click between the rails. This will be our 'fixed' piece of track - the second piece we click will move to join it. Click one end of another piece of straight and note that it 'jumps' to join up with the first. Join the third piece to the other two, and you should now have a long line roughly around the middle of the layout

Now add two very simple stations to the ends of the stretch of single line. From the Track palette, add a right and a left-hand point, one near each end of the straight, and connect them up to the straight using the Joining Tool. Take care to click the straight first, and add the points the right way round. We want to arrive at the station and have the choice of two platforms, so the toe end of each point is to be joined to the straight.

Now add platforms. You may find this easier if you zoom in. Click the 'Zoom Rectangle' button and click/drag the mouse over the left-hand six feet or so of the layout. Add another length of straight and join it to the 'straight on' road of the point. It may run over the edge of the layout. Use the Move tool to position the track (click and drag the mouse) so that the straight runs along the centre of the layout, and ends just a little way short of the layout edge.

Using Flexible Track and the Spacer

We now wish to add a second platform, parallel to the first, and joined to the diverging road of the point with an appropriate curve. This is where the Spacer tool and the Flexible Track part come in handy. The Spacer tool is a dummy track part. It doesn't actually contain any track, but it does have two ends, whose relative position is fixed. You can change this distance by right-clicking on the spacer tool and bringing up the 'Properties' box. This will allow you to change the interval between the two tracks to be spaced. It is set by default to the standard spacing between adjacent tracks for the scale and gauge of your layout. Place a Spacer tool, for the moment anywhere on the layout.

Select the Joining tool, then click the end of the platform road and attach the appropriate side of the Spacer, leaving the other arrow where you want to position the second platform. Position a length of Flexible track roughly parallel to the platform, then join it to the spacer using the Joining tool (figure 7.3). Join its other end to the diverging road of the point. Note that the dotted lines of the flexible track have been replaced by a straight and a curve of the radius required to fill the gap. You can use flexible track in this way to join any two track ends. *Trax* will attempt to find a configuration of straights and curves that will do the job. Once it has done its work, right-click and delete the spacer tool.

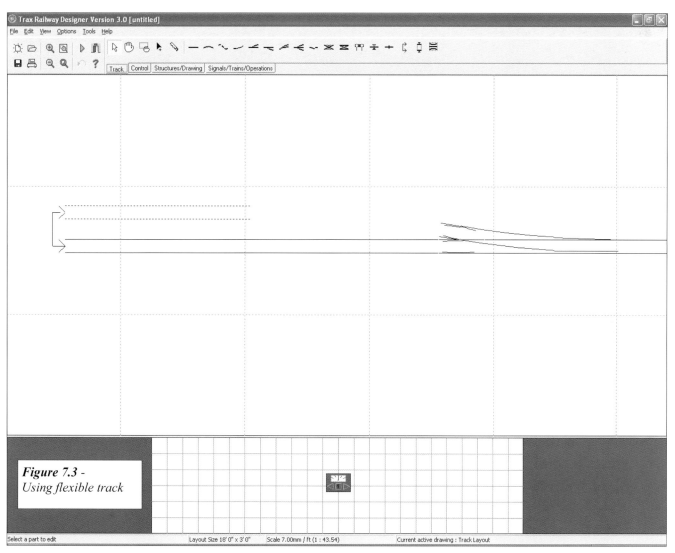

*Figure 7.3 -
Using flexible track*

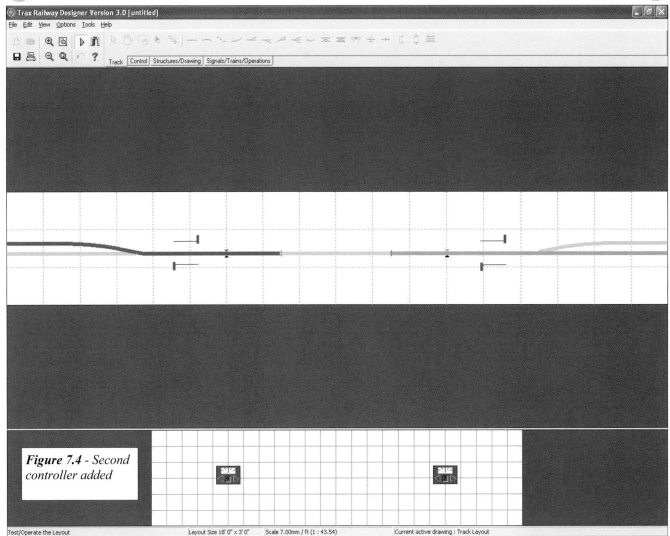

Figure 7.4 - Second controller added

Repeat the process at the other end of the line. Use Ctrl+Right Arrow to get there. Take care over which way the spacer faces. Remember we are making a mirror image of the first station, so the arrows this time need to face inwards. Return to Full View (Ctrl+Home). You should now have a long single track main line with two very simple stations, which we will call A and B, at either end.

Click the 'Test/Operate' button to go into test mode, and you will find that by clicking the points you can change their setting. Exit test mode (click 'Test/Operate' again).

Signalling

Now let us add some signals. Zoom back in to the left hand station (A), click the 'Signals' palette, and select a signal. Position this above the main line, beyond the point. This will be our starter signal. Then position a second signal opposite it, to represent the home signal. You will find this is the wrong way round, so right click it, go into Properties and change the orientation from West to East.

Adding Track Feeds and Breaks

We now want to add some power, so go to the 'Controls' palette. The white rectangle below the layout is the control panel, and this is where we position our controls. *Trax*3 will automatically position a train controller on the control panel, it being assumed that every layout will need at least one. Position a second controller to the right. Note that controllers are automatically colour-coded. Place a Track Feed from the Track palette on the straight between the two signals. Right-click it, then in Properties put 'perm' in the 'Connect to controller 1' box. This specifies a permanent connection.

Now click the Test/Operate button and note that we have red colour-coded track where there is power, and grey where there is not (in the platform road not selected by the point.) However, if you zoom out to a full view, you will see that the red colour-coding extends right into station area B, which is not necessarily what we want. Therefore, we need a break in the track just beyond the feed. Exit Test/Operate mode (you cannot add track parts in this mode) and place a Track Break just beyond the feed. Test again, and you will see that we now have limited the scope of the red controller to station area A only.

Next, zoom in on station B, and repeat the process of adding signals, track feed and break there. Connect the track feed to controller 2 this time, and when you now test the layout in Full View, you should have a red and a green colour-coded scope for the two controllers with a grey dead section of line between them (figure 7.4). Clicking the points and signals in Test/Operate mode will change them.

Operating from a Lever Frame

All our signals and points at present are manually operated. We will now automate them, and in the process add some electrical switching to feed power to the 'dead' track at appropriate times. We will adopt the principle that operators drive trains towards themselves, and use the layout signalling to decide which operator that should be.

Signals and points will be operated by a lever frame, so go to the Controls palette, and select the Lever Frame button. Place a frame above each of the controllers on the control panel, then right-click it, select Properties, and change the number of levers to 3. We only need a lever for the point and two for the signals at each end. You can have up to 100 levers on a single frame if you wish.

Now we will assign levers to points and signals. Right click the point in station A and in its Properties dialog, in the 'Setting changed by' box, enter L1. Levers in the first lever frame are called L1, L2, etc. Those in the second, K1, K2 and so on. Go back into test mode and observe that lever L1 now changes our point. You can no longer change it manually, ie by clicking on it. Points and signals are either manual or automated, never both. Assign lever L2 to the outgoing starter signal, and lever L3 to the incoming home signal in the same way. Test that these operate in the way you would expect, then repeat the process for station B, taking care to assign lever K2 to the outgoing signal.

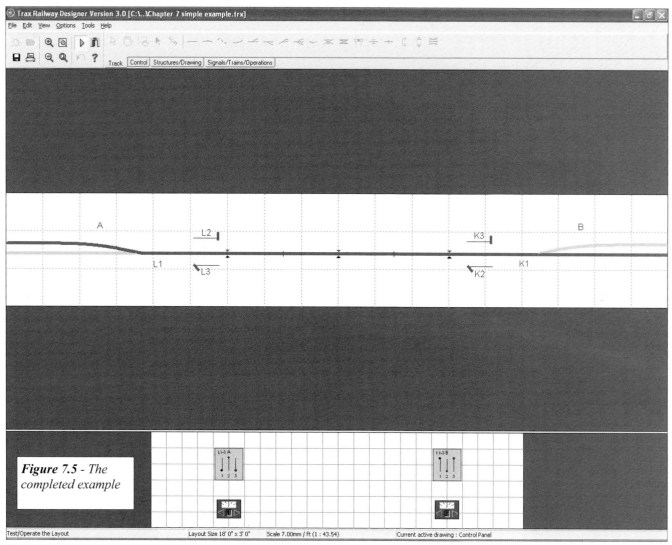

Figure 7.5 - *The completed example*

Labelling

You will find it easier to operate the layout if you label the points and signals, so go to the 'Structures/Drawing' palette, select the Text Label button, and place a label adjacent to A's starter signal. Right-click the label, and in its Properties dialog, give it the caption 'L2' (because it is operated by lever L2). Set its size so that it is easily legible. Note that the size of labels is given in mm, as if they were actually painted on the layout. This is more convenient than using point size, as a legible label at most scales would need to be several hundred points in size. You can adjust the position of the labels using the Move tool. Label the other signals and points with their appropriate lever numbers.

Connecting Power to the Track

Now we wish to feed some power to the central dead section. Position a feed here, go into its Properties dialog, and under 'Connect to controller 1 if' type 'L3'. Similarly 'Connect it to controller 2 if' type 'K3'. If you now go into Test/Operate mode, you should find that each of the 'incoming' signal levers cause the central section to be assigned to the controller whose home signal is off. If you reverse both home signals, you will get a message from *Trax*, indicating that it has detected an error - the central feed is connected to two controllers, which is not advisable! We will use lever locking later to prevent this combination of levers.

There is still one more step to take. We have a switchable central section, but if we want our two controllers to drive a train all the way from the opposite station, we also need to extend the scope into that station, not just the central section. Therefore, we will introduce a further operating rule for the feeds at the station throats. Right-click the feed in station A, whose connection is at present permanently to controller 1, and change this to controller 1 if '-L2'. The '-' sign is short-hand for NOT, so the feed will be connected to controller 1 unless L2 is pulled. Then, under 'Connect to controller 2' enter the rule 'L2&K3'. Having done this, repeat the process for the feed in station B, assigning it to controller 1 if 'K2&L3' and to controller 2 if '-K2'.

Next, open the Properties dialog for the two lever frames, and add the rule that lever 2 locks lever 3, and vice versa. This will prevent the short circuit referred to above. (To be fully authentic, of course, we should have a pair of signals for each platform, i.e. a total of four at either end. If you wish, you can amend the layout accordingly.)

If you now go into Test/Operate, you should find that the red controller has control of station A, unless the Starter signal (L2) is off. The green controller has control of station B unless K2 is off. When both starter and home signal for a given route are both off, control is with the controller in

whose station the train will stop. Once the train has passed the starter signal, it can be returned to danger, and the 'local' controller regains control over the station, whilst the other controller retains the central section. Thus shunting could proceed whilst the train makes its way over to the other station. Figure 7.5 shows the completed layout.

This is an extremely simple example, but it illustrates the process of using *Trax*. The same process would enable you to build up a large layout with multiple controllers and a complex track formation, then test out your wiring logic. Having worked through this example, you could now look at some of the other sample layout files included with your *Trax* installation.

Using Switches and Relays

In addition to the Lever Frame component, *Trax* has a number of switches which you may find useful if you want to design an 'OCS' or 'NX' style control panel as described earlier. The toggle switch can be referenced in operating rules for points, signals, and other components exactly as a lever. For example, 'SW1' is true if SW1 is in the 'down' or 'on' position; '-SW1' is true if it is 'off'. In the case of 3-way switches, '-SW2' is true when the switch is in the centre-off position, '>SW2' is true if it is 'on' in the 'up' position, '<SW2' is true if it is 'on' in the 'down' position.

Push-buttons differ from toggle switches in that they are momentary in action. That is, they are permanently in the off position except when the mouse pointer is over them and the mouse button clicked. Track and other parts operated by such a button require a special format of operating rule. This generally consists of one push-button to put them in state *a*, followed by a semicolon then a second button to put them in state *b*. For example, a set of points with an operating rule 'PB1;PB2,PB3' would be reversed by PB1, and set normal by either PB2 or PB3.

The relay component can be useful to test your relay logic. *Trax* does not do electronic circuit simulation, but the relay component can be used to illustrate how the circuit works. As an example, there is a *Trax* file 'Abbotstone NX FY.trx' in your samples folder which illustrates the circuit given in figures 6.11 and 6.12.

'Read Only' Properties

If you add a Track Part to your layout, for example a point, then you will normally right-click it and select the switch blade length, crossing angle etc. appropriate to your needs. However, once you have joined this part to any others these properties become read only. You cannot alter them with the Properties box. If you need to adjust a part already joined to others, right-click it, then select 'Isolate Part'. You can then move it or change it, and rejoin it to its neighbours in the sequence you require. Properties that do not affect the geometry of adjoining parts can be altered even when a part is connected up.

Lost Curves

If you carry out the above procedure on a curve, you can 'lose' it. Basically, unattached curves are always drawn symmetrically about the 'twelve o'clock' position. Thus, for example, a 60-degree curve would extend from the '11 o'clock' position to the '1 o'clock' position. When you first create a curve, it will be created in this way with the left-hand end where you clicked. If you subsequently attach it to other track parts, it may end up in some completely different orientation. Subsequently, if you Isolate it and change its length or radius, *Trax* will re-draw it in the '12 o'clock' position, keeping the same centre point. This may take it off the layout. To find it again, simple zoom out (Ctrl+PgDn) until you see it, then use the Move Tool to bring it back onto the layout.

Short Lengths of Track

If you use a length of flexible track to join two ends, one of which is a spacer tool you may on occasion have difficulty trying to join something else onto the end from which you subsequently delete the spacer. *Trax* will give you an error message saying the end is already joined, when apparently there is nothing attached to it. Invariably this turns out to be a very short length of track, sometimes so short it is not at all visible on the track plan. Where you think you are clicking on an unjoined end, you are in fact selecting the end of a track part already attached to the very short length. The problem is solved if you use the Parts List tool. Select 'Short Lengths' then click 'Update' and *Trax* will list all lengths less than 1" or 25mm. You can then click on these, press 'Select', and they will be highlighted on the layout track plan. Using the Edit menu, you can look at their Properties, Isolate or Delete them.

Trax will try to avoid short lengths, preferring to increase the length of an existing straight instead. However, when using the spacer tool, this option may not be available to it.

Locating feeds and breaks

Every track part has the facility to take a track break at each of its ends, plus one feed (from up to four controllers). The feed is normally in the centre of the part, except for points, where it goes at the toe end. When you click a part to add a break or feed, it will appear in one of these fixed positions. If the positioning of your track break is critical, you should adjust the length of the track part onto which the break is added.

Selecting Awkward Parts

Sometimes, it is hard to click on the part you want. Try zooming in - this always makes things easier. However, you may still have difficulty, for example, if you are trying to click on a track break between the diverging roads of two points in a crossover. In this case, use the Parts List. Simply double click the name or ID number of the part you want to select and click 'Select Part'. You can then use the Edit menu to change its properties, isolate it, or delete it.

Trax Templates

Trax has the facility to print actual-size templates to allow you to build your own track. In using these templates, note that the line drawn for the rails represents the *inner edge* of the running rail. *Trax* does not know what kind of rail you are going to use to build your track. All it knows is the track gauge, which is the distance between the inner faces.

The geometry of points and sleeper spacing are compromises. Each railway company had its own permanent way department, and although they all worked to the same track gauge and general principles of construction there were differences in details. *Trax* does not attempt to reproduce any one railway's trackwork precisely. Rather, it uses a set of fairly typical rules to space sleepers and other features so that the track looks reasonably representative of virtually any standard gauge railway.

Printing Templates

When printing templates, *Trax* takes a slightly different approach from the commercially produced templates that are available. In effect, the template is simply a zoomed-in view of a part of the layout. It may encompass one or more individual track parts, such as points or slips, or in the larger scales it may only be possible to fit a part of a point on one sheet of paper.

This approach is taken because, when points and crossings are built adjacent to each other, their sleepers may interfere with each other. As with the prototype, some judgement must be exercised by the builder. Maybe you would choose to move the sleeper positions slightly to allow them to interlace, or you might prefer to replace all of them by sleepers going straight across both lines.

To print the template, we need first of all to set up our printer in either landscape or portrait mode, as preferred, and select 'Show Sleepers' from the Options menu. Then, on the 'Structures/Drawing' toolbar, we use the page size tool to place a rectangle on the track plan which represents the printable area of a sheet of paper (this is generally smaller than the paper size because of your printer's need to

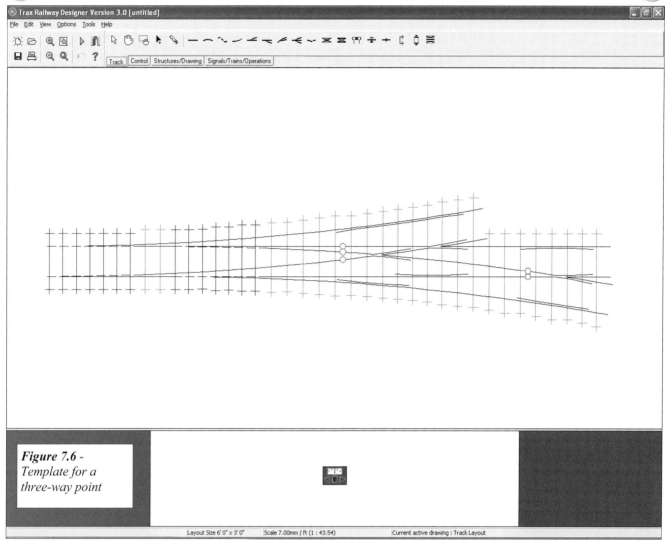

Figure 7.6 - *Template for a three-way point*

hold the paper whilst printing). You can move this around with the move tool until it is in the correct location. Further pages can be added, until the required area of the layout is covered. Then, simply press the print button or select the File/Print menu item, to bring up the print dialog box and tick the Templates option.

Trax will then print an actual size template for the track part (s) selected. To work out the size, *Trax* uses your printer driver software to find out how many dots represent one inch, and scales the picture accordingly. However, a word of caution is appropriate. Not all printers have absolutely accurate registration, and sometimes the dots per inch figures are rather more nominal than might be hoped. Also, like many drawing programs, *Trax* approximates curves by drawing a number of successive straight lines. Therefore, the template you produce from *Trax* is a *guide only*. Do not use it to line up rails by eye. An accurate track gauge is essential equipment, and this should be used for actually positioning your rail.

On the template (figure 7.6), sleepers are represented by a long line, which is their centre-line, and two short lines at right angles, which represent the ends of the sleeper. They are aligned with the 'straight on' direction for points. Some railways, notably the Great Western, aligned sleepers at right angles to the arc through the central crossing vee. Again, if you wish to follow your chosen prototype's practice exactly, you will need to research this point. The locations of necessary track breaks are indicated by circles.

Creating Custom Track Parts

Trax has built-in capability to calculate the geometry of a wide range of track parts such as left and right hand points, Y points, curved points, three way points, diamond crossings, single and double slips and catch points. However, there may be a requirement on your layout for a configuration that is not included. For example, if you wanted to build a double track junction on a curve, you would need a diamond crossing in which both legs were curved to different radii. *Trax* only draws diamonds with straight legs.

For help with producing such track parts, the alignment tool will be found invaluable. The alignment tool can be used to align the ends of curves and straights with one another. For example, you might want a three-way point with all three roads curved. This is not a supported part in *Trax* – the 'standard' three-way has one road straight. However, by positioning the alignment tool at the 'toe' end and 'joining' three curves to it, you can simulate an all-curved three-way point.

When you create a template in this way, note that it contains only the running faces of the main rails. The builder will have to pencil in the appropriate check rails and wing rails, and work out sleeper locations etc. However, once a little experience has been obtained in building standard track parts, these should present no problem.

When you use the alignment tool in this way, you may find it difficult to select the aligner for joining up rather than a length of track already joined to it. It can be done, if you click within the aligner rectangle, but outside the gauge of the track. However, unless you are zoomed right in, this is not easy. *Trax* will bring up the Multiple Parts Selected dialog box in such cases, showing all the parts which it thinks you might mean to indicate. Select the appropriate one by using the up and down arrow keys, then hit return.

The Tools Menu

Under the Tools menu item, you will find a variety of useful tools which will be of value during track design.

Conversion Utility

This allows you to enter a measurement in either metres, millimetres or feet and inches, and to convert it. *Trax* will work out from what you have typed which conversion is required. If you enter a figure with no units, *Trax* assumes these are millimetres.

Below the edit box are three buttons. The first button does a straight conversion. Thus if you type 1" in the edit box, the result will be 25.40mm. If you type 25.4mm the result will be given as 1.000 inches. The second button scales down the measurement you have typed according to the layout scale (note that for this reason you cannot use the conversion utility until a layout has been set up). The third button scales the other way. The measurement you type in is assumed to be an actual measurement on the model, and if you click the 'scale up' button, the result will tell you what this measurement on the model represents in real life.

Minimum Radius Calculator

There are several different interpretations of 'minimum radius'. The absolute minimum, below which a vehicle could not even be placed on the track with its tread touching the rail head, is not really achievable in practice. The calculator allows you to see the effect of errors in wheel alignment, track gauge, etc. Normally, you should allow at least 1% tolerance, and this will give you a more realistic radius that might be usable.

At the other end of the scale, the minimum radius that your vehicles could negotiate without contact between flange and rail running face is generally much too large to work to in any practical layout.

It should be understood that in the context of minimum radius calculations, what counts is the *fixed* wheelbase. For example a bogie coach with two 7ft wheelbase bogies, 43 ft apart has a fixed wheelbase of 7ft, not 50ft as might be thought. Similarly, for a 2-6-4 locomotive what is important is the wheelbase of the central 6 wheels. Because the bogie and pony truck are able to move laterally, they will not restrict the radius the engine can negotiate.

Clearance Calculator

With long bogie coaches on tight curves, however, a second factor comes into play. This is the overhang between the bogies on the inside of the curve, and the overhang of the ends of the vehicle on the outside of the curve. When two bogie coaches pass each other on a curve, these overhangs can result in a collision if the tracks are too close together.

The standard distance between tracks, called the *interval*, is one of the parameters you set up when you start a new layout. It is used, among other things, to determine the length of turnout curves on points and crossings so that two identical points joined back-to-back will give exactly the right separation to form a crossover.

This distance on the prototype is generally just over 11 feet. Prototype vehicles are rarely more than 9 ft 6 ins wide, so that there is ample clearance between vehicles on straight track. However, on a curve the overhangs come into play. If the curve is tight, or if the vehicle is extra long or wide, then the overhangs may reach so far that they exceed the standard interval between tracks.

The clearance calculator will alert you to this possibility. You may then wish to increase the separation of tracks on the more tightly curved sections of the layout to avoid the possibility of collisions.

Curve Calculator

When two curves of different radius are joined, they must have a common tangent at the junction point. This is only achieved if the junction and their respective centres of cur-

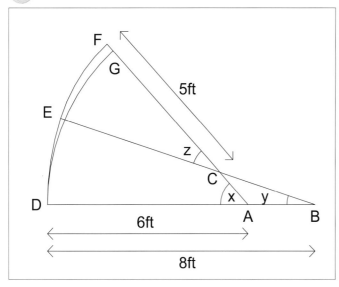

Figure 7.7 *- Curved point geometry*

vature lie on the same straight line. When designing curved points, you must ensure that this condition is met. The curve calculator tool enables you to do this.

As an example, let us design a right-hand curved point to take a single line of 6ft radius out to a pair of parallel lines one standard interval apart. One way of doing this would be to set up a curved point that looked 'about right' then join it up to an appropriately sized spacer using flexible track. However, such an approach could lead to a hotch-potch of straights and curves of different radii. Far better to design it with the minimum number of changes of radius using the curve calculator.

Referring to the diagram (figure 7.7), we have illustrated an '0' gauge curved right-hand point. For clarity, only the centre lines of the track are shown. The inner track, running from point D to point G is of radius 6ft, and its centre is at point A. The outer track, of radius 8ft, starts at the same location, D, and has its centre at B. As we move round the curve, these two tracks diverge. At some location, labelled E in the diagram, we need to change the radius of the outer track from one that is *larger* than the inner radius to one that is *smaller*, so that after a while the 'new' outer arc and our original 6ft curve will be both exactly one standard interval apart, and exactly parallel. We have labelled this location F.

Suppose we choose 5ft as this smaller radius. Our problem boils down to this: exactly where should we locate the change of radius in the outer curve (i.e. point E) so that when the tracks draw parallel they are exactly one interval apart? In other words, we need the angles labelled x, y and z.

Because of the requirement for a common tangent at changes of radius, we know that points D, A and B must lie on the same line, and likewise points E, C and B. We also require the exits to be parallel so points A, C and F must be in line as well. Now, if you happen to remember enough of your school geometry, this gives you the necessary information to solve triangle ABC and thus work out the angles x, y and z. Alternatively, you can simply type the three radii into the curve calculator, press the calculate button, and read off the answers! As we see in figure 7.8, these are 47.584°, 18.100° and 29.484 respectively.

You can now set up a curved point using the curved point tool on the Track palette, then right-click to bring up its properties box. Selecting the 'curved point' tab, set radii 6ft and 8ft, uncheck the 'end curve at check rail' boxes and instead continue the curves through angles 47.584° and 18.100° respectively. This will give you a curved point as required.

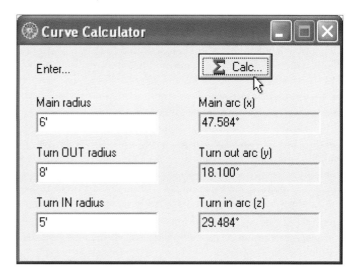

Figure 7.8 *- Curve calculator*

8

TRAX 3

We will now turn to those features of *Trax 3* that are new. Some of these are in response to various requests that people have made, others are to extend the capabilities of *Trax* into the areas covered by this book, namely signalling and lever frames.

What's New?

Trax 3 includes one or two new options on the 'Layout Settings' form. When you start a new layout, you can opt for a different setting for clearance lines either side of the track instead of the standard clearance for the modelling scale you select. You can also opt for a different standard sleeper length for your track templates. On the 'Track' toolbar, there are a couple of new buttons. The 'Transition Curve' button lets you create a length of track which starts straight, then gradually increases its curvature to a given final radius. The properties box for the transition curve allows you to specify the overall length and the final radius. You can then join this to a straight and a curve of the specified radius to make a realistic transition curve. A second new button is the 'Scissors Crossing' button. This gives you a complete scissors crossing. You can specify the crossing angle and the track spacing between centres, and *Trax* does all the rest!

As well as the ability to move track, including joined track parts, *Trax 3* now lets you move whole sections of the layout, buildings and all, to a new position using the 'Window Move' tool. This can be a great time-saver if you wish to change the overall size of the layout. On the subject of buildings, *Trax 3* has a lot more features to represent buildings and other scenic features. A new toolbar called 'Structures/Drawing' gives you buttons to represent rectangular buildings (this was all you had in *Trax 1* and *2*), filled circles ('trees'), gradient symbols to represent embankments or cuttings, then straight lines, perpendicular or parallel lines, polygons (filled or outline only), outline circles and rectangles.

On the 'Control' toolbar, a couple of new buttons let you add light emitting diodes (LEDs) and relays to your layout or control panel. The relay component can provide a useful visual representation of whatever relays you have in your electrical equipment, and being able to assign them operating rules means that you can test out your relay logic visually. The CD contains some examples of this. The relay component can also simplify a lot of your operating rules. Suppose you had an operating rule such as 'L71&-L70&-L69&L66' (see table 4.4). This rule is used in the Itchen

Bottom layout to operate signal 71a, also as part of the operating rules for four separate track feeds. You would thus need to type it in five times. However, by assigning it to a relay called, say, 'R71a' you could just type this as the operating rule for your signal and track feeds.

There is a brand new toolbar 'Signals/Trains/Operations' which is where the buttons relating to the most important advances in *Trax 3* are located. These are a greatly increased facility for signalling, which now includes full interlocking, and a facility to simulate the operational running of your railway and, whilst doing so, to collect a script which you can then use to produce operator cue cards for your operators. It is these features upon which we shall concentrate shortly.

Also included on the CD are two smaller programs which you might find useful if you are building a lever frame. One checks out your lever frames and comes up with a list of all the valid combinations of lever pulls your locking rules will allow. These can then be tested using *Trax 3* to ensure that you haven't missed anything. The second little program is provided to help you with the mechanical design of a set of locking bars. It can take a set of locking rules from a *Trax* file and help you turn these into a set of locking bars, along the lines discussed in chapter 5.

Finally, the *Trax* Help file has been converted to the latest HTML Help format. Microsoft Corporation no longer offer support for the old .HLP file format, and with Windows 7 they apparently no longer even provide a reader for it. Therefore, *Trax 3* Help has been upgraded.

Lever Frame Enhancements

The Lever Frame properties box (figure 8.1) has been greatly upgraded from that which was offered in *Trax 2*. The principal enhancement is the introduction of a grid on which you can specify, for each lever on the frame, a description, a colour, a sequence of other levers locked by this lever and a sequence of other levers released by this lever.

When typing in your lever descriptions, bear in mind that if you are developing a script (see below) *Trax 3* will add a line to the script for the operator designated for this signal along the lines of: 'Wait for …mydescription… signal'. Therefore, do not include the word signal in your description, otherwise you will get an instruction that says 'Wait for down home signal signal' or something similar!

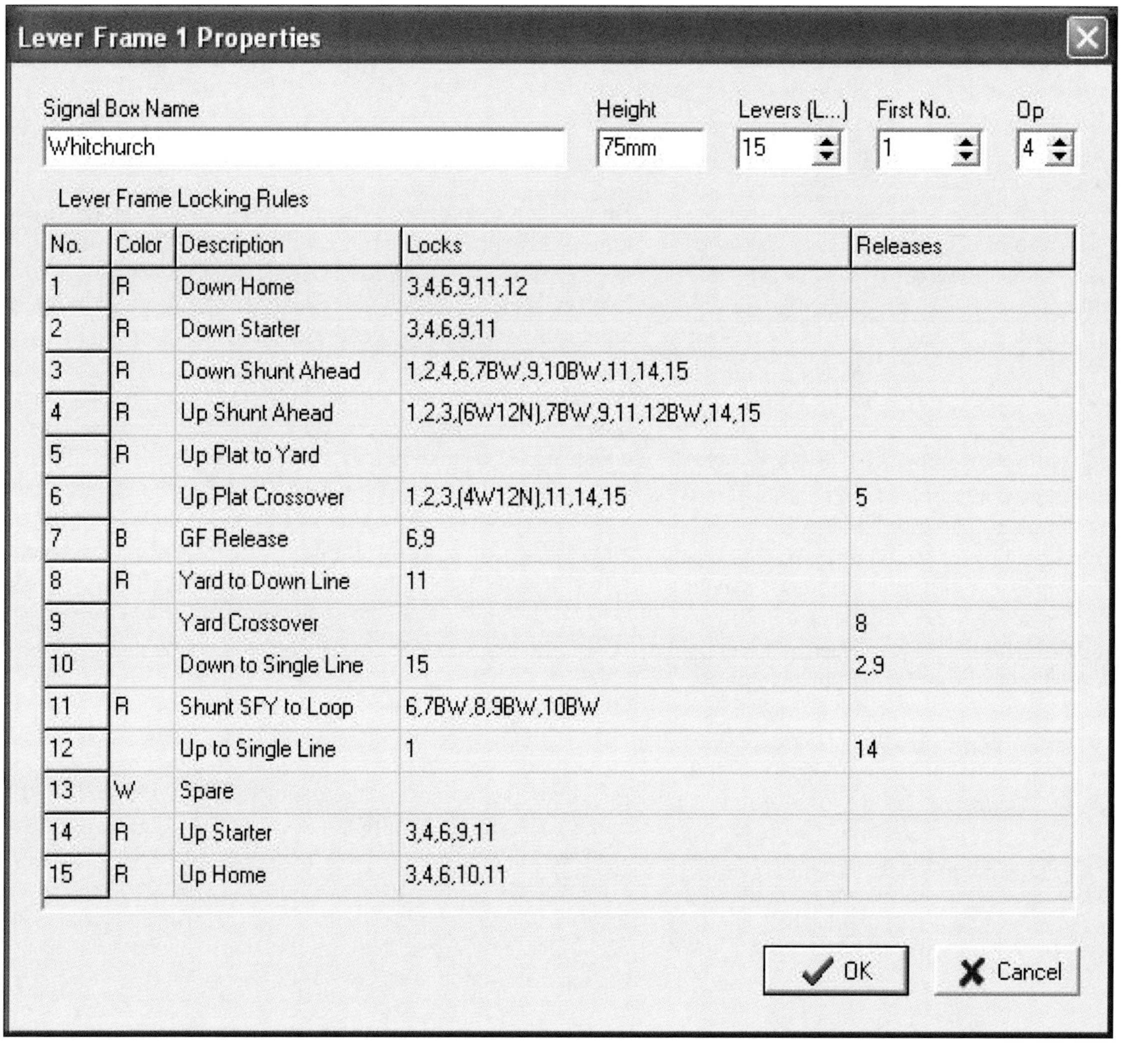

Lever Frame 1 Properties ☒

Signal Box Name | Height | Levers (L...) | First No. | Op

Whitchurch | 75mm | 15 ⬍ | 1 ⬍ | 4 ⬍

Lever Frame Locking Rules

No.	Color	Description	Locks	Releases
1	R	Down Home	3,4,6,9,11,12	
2	R	Down Starter	3,4,6,9,11	
3	R	Down Shunt Ahead	1,2,4,6,7BW,9,10BW,11,14,15	
4	R	Up Shunt Ahead	1,2,3,(6W12N),7BW,9,11,12BW,14,15	
5	R	Up Plat to Yard		
6		Up Plat Crossover	1,2,3,(4W12N),11,14,15	5
7	B	GF Release	6,9	
8	R	Yard to Down Line	11	
9		Yard Crossover		8
10		Down to Single Line	15	2,9
11	R	Shunt SFY to Loop	6,7BW,8,9BW,10BW	
12		Up to Single Line	1	14
13	W	Spare		
14	R	Up Starter	3,4,6,9,11	
15	R	Up Home	3,4,6,10,11	

✓ OK   ✗ Cancel

Figure 8.1 -
Lever frame
properties

Levers will be coloured black unless you specify something else. Distant signals are yellow, specified by the letter 'Y' in the colour column. Stop signals, whether running signals, subsidiary signals or ground discs, should have red levers, specified by 'R'. Locking bars, ground frame releases, and the like are generally blue, enter 'B'. Level crossing gates, and some other release levers were brown; use 'N' for these. King levers, and distant signal levers in the early days, were green, use 'G' for these. Finally, if you have any spare levers (and it is often a good idea to design your frame with at least one, in case a future need crops up) colour these white using 'W'.

In *Trax 2*, each lever frame started with lever number 1 and went up sequentially to the total number of levers. To distinguish levers on different frames, in operating rules for example, those on the first frame added to the layout were designated L1, L2, L3 etc. Those on the second were K1, K2, K3... and so on, using the letters J and M for the third and fourth frames. For compatibility purposes, *Trax 3* recognizes these designations, so if you already have a layout designed with *Trax 2*, you can move it into *Trax 3* without

redesignating all your levers.

However, *Trax 3* also offers an alternative. Instead of having levers designated by different prefix letters, you can distinguish levers on different frames by having number sequences that can start with numbers greater than 1. For example, your first lever frame can have levers 1-15, your second can have levers 16-24, and so on. It is less ambiguous than the first scheme, particularly when you have a number of operators, and you want one of them to pull lever 10, say. If they all have a lever 10, confusion could result. By numbering levers differently, you make each lever number unique and thus unambiguous.

To use this facility, enter the number of levers on your frame in the 'Levers' box and the number of the first lever in the 'First No.' box. Although *Trax* does not specifically prevent you mixing the two numbering schemes, it is strongly recommended that you keep to one system or the other. If you opt for the second scheme, note that all levers will have frame letter 'L', whichever frame they are on. Thus levers on the first frame are designated L1 to L15, those on the second L16 to L24 etc.

The 'Op' box lets you specify an operator for this frame. When you go into Test/Operate mode, each time you reverse or restore a lever, an appropriate instruction will be added to the script for this operator.

Locks and Releases

These two columns in the lever frame properties box are the key to making your layout operate like the real thing. They provide the facilities necessary to emulate route/signal inter-locking and absolute block working as on the prototype.

In the 'Locks' column, enter a list of all the levers which are to be locked when this lever is reversed. Enter the lever numbers separated by commas. If the lock is to be a 'both ways' lock, enter 'BW' after the lever number. To lock a lever conditionally, place the lever number to be locked, followed by 'W' (for 'when') then the lever number to which the condition applies, followed by either 'N', meaning apply the lock when this lever is normal, or 'R', meaning apply it when the lever is reversed, in parentheses. For example, in the row corresponding to lever 1 you could enter '(6W2N)' in the locks column, meaning that lever 1 locks 6 when 2 is normal.

You do not need to fully specify mutual locks. If you tell *Trax 3* that lever 1 locks lever 12, you do not need to also tell it that 12 locks 1, although you can if you wish add it for clarity. *Trax* will work these things out for itself, however.

You can lock a lever on another frame. To do this, simply specify the frame letter in the locking rule. For example, suppose in the row corresponding to lever 1 on your first lever frame (L1) you entered in the locks column '3,4,K6'. This would indicate to *Trax 3* that you wanted lever 1 to lock levers 3 and 4 on this frame, and also lever 6 on the second frame. Note that you must include the frame letter if the lever is on a different frame, even if you have opted for the scheme whereby all levers have letter 'L'. Inclusion of the frame letter indicates to *Trax 3* that this lever is on an-other frame, otherwise it thinks you have simply mis-typed a number that is outside the range of lever numbers for this frame.

In the 'Releases' column, enter the numbers of those levers which are to be released when this lever is reversed. As with locks, your releases can apply to levers in different frames. A typical use of this facility would be to have a main frame and a ground frame. A lever in the main frame would be a 'ground frame release' lever, which would re-lease a 'king' lever in the ground frame. This king lever would then in turn release the other levers in the ground frame. It would also lock the ground frame release lever in the reversed position. As before, to specify a lever on an-other frame, precede the number with the appropriate letter.

Like locks, releases can be conditional. If you entered a rule such as 1 releases (2W12R), this would mean that 1 released 2 when 12 is reversed. With 12 normal, 2 can be operated independently of 1. However, when 12 is reversed, 2 can only be pulled if 1 is reversed first. The 'BW' suffix does not apply to releases. If you think about it, having one lever release another 'both ways' is a nonsense. If you want to put in sequential locking, as described in chapter 5 then you have to use what looks like a rather bizarre format. For ex-

Figure 8.2 - Script window

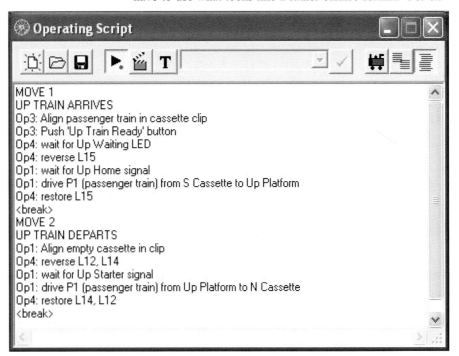

ample, suppose we wanted to arrange matters so that lever 1 could only be pulled when lever 2 was normal (see figure 5.10). We would enter this into Trax as a release rule 2 releases (1W2N). It looks a bit odd, but Trax 3 knows what it means.

Scripting

We now turn our attention to the other major enhancement in *Trax 3*, the introduction of scripting. If you want to operate your railway to a timetable or schedule, and you have more than one operator, then it is essential that everyone keeps in step. The traditional way to ensure this is for each operator to have a set of 'cue cards'. These are generally the standard type of card used in a card index, upon which are written the actions that the operator is to carry out for one particular move.

For example, you might have operator A prepare a train on a cassette in the fiddle yard, then operator B, who is a signalman, might set the appropriate route and pull off the necessary signals. Finally operator C would be called upon to drive the train to its destination. Once this move is complete, all three operators turn over their cards for the next move, and so on.

Trax 3, through its scripting facility, allows you to automatically capture the data from which you can prepare these cards. It also allows you to test the timetable on the computer to make sure, for example, that all the trains are back in their starting positions for the next run through the timetable, and that you have sufficient storage cassettes or sidings in your fiddle yard to accommodate the trains that arrive.

The scripting tool is invoked by the Tools|Script menu item or by pressing the F2 function key. The scripting tool window (figure 8.2) incorporates a small toolbar at the top which includes the usual buttons to start a new script, open an existing script or save the current script. The next group of three buttons comprise a 'record' button, by means of which you can turn recording on or off, a 'clapperboard' button, which you use to end a move and start the next one, and a text button by means of which you can enter plain text into the script. The script itself is in a standard edit box, so you could just as well type it here. However, the benefit of using the text button is that *Trax 3* remembers what you type. Therefore, if you have a phrase you will use a lot, for example 'down train arrives', you can then repeat it with a couple of clicks rather than typing it again. The three buttons at the right of the screen give you the ability to add a 'snapshot' of where all your trains are to the script, and finally to sort the script into either time sequence, or grouped by operator. Use time sequence to develop and test a script on your PC, use operator sequence to prepare your cue cards.

Operators

New to *Trax 3* is a field called 'Operator', generally shortened to 'Op'. This is present on all *Trax* parts that can be 'clicked' in Test/Operate mode such as levers in a lever frame, push-buttons, toggle switches etc. It is also present on parts that can be altered by means of such clicks, for example signals and LEDs. In the first group, the 'Op' field is interpreted as the operator who is to be instructed to carry out this action, for example 'Op1: Reverse L1' is an instruction to operator 1 (the signalman) to reverse lever L1. In general, components on the control panel will generate this type of message.

With signals, and in some cases LEDs also, the idea is to indicate to the appropriate operator that they are to wait until the signal comes 'off' or the LED comes 'on' before starting a train or carrying out some other action. In the above example, if we had assigned lever L1 to signal 'down home', and assigned this signal to operator 2, then with a single click of lever L1 we would generate in the script two lines: 'Op1: Reverse L1' and 'Op2: Wait for down home signal'.

Trax 3 will collect all of your actions in the sequence in which you click the components. To prepare the cue cards, you need these in operator sequence. Clicking the 'Sort by Operator' button will achieve this. Use the Clapperboard tool to break your timetable up into moves. A single click will end a move and insert the word '<break>' into the text. Clicking a second time will add a header to the script indicating 'MOVE n' where n is the next move number in sequence. When you sort by operator, *Trax 3* will list all the moves, starting with move 1, for operator 1, then all the moves in sequence for operator 2, and so on. You can have up to 9 operators.

Trains and Zones

Trax 3 enables you to place 'trains' on your layout. These are not representations of trains as such, but resemble more those little yellow 'Post-It' notes which you use to remind you of things. Their purpose is to let you know where your trains are at any time. They can be moved around the layout during the operating session and they can also be split up or combined. The trains have a tag which is a short mnemonic, for example 'P1': this appears on the 'Post-It' note. They can also have a lengthier description such as 'Passenger Train 1' which will appear in the script.

The trains are placed on the layout by means of the 'Train' button on the 'Signals/Trains/Operations' toolbar. This enables you to give your train a tag, a description and a colour coded 'blob' if you so wish. When in test/operate mode, you can then 'drive' these trains around the layout by clicking and dragging with the mouse.

To indicate in the script where trains are to be driven to and from, *Trax 3* enables you to define zones of your layout. These are rectangular areas which might cover a single length of track, such as a platform, or a whole area, such as a goods yard. You can also define fiddle yards as zones, or give zones descriptive names such as 'up platform' or 'clear of crossover points'.

In addition to driving a train from one zone to another, you might wish to split it, for example during shunting. To do this, right-click the train then click 'Split' on the Train Properties dialog box, and you will be given the opportunity to create a second tag and description. You might make this the goods engine, in which case you might change the original tag to 'goods wagons'. An entry will be created in your script to indicate that the goods train has been split into goods engine and goods wagons. You can now manipulate these entities separately.

You can merge the two back together simply by dragging and dropping the engine onto the goods wagons. You will be presented with another dialog box asking for a name for the combination, and a line such as 'attach goods engine to wagons to form goods train' will be added to the script.

The final button on the Signals/Trains/Operations palette is an 'Operator Action'. This places a component on your layout which looks like a push-button. However, instead of interacting with other components via their operating rules, this button simply places a line of text in the script. Use this to include actions in your script which are not covered by the standard *Trax* mouse clicks. For example, you could have a button which said 'Op1 : set up goods train on storage siding 1'.

Timetable Testing and Editing

You can save your script to a text file, then test it by turning off the 'record' button, and following through your operator actions. You will need to have them in 'time sequence' to do this. Once you are happy that the timetable works, and that the trains end up in their correct locations after one timetable cycle, you can sort them into 'operator sequence'.

The *Trax 3* Operating Script dialog box has only very rudimentary facilities for editing and formatting. It is recommended that in order to produce cue cards you save the file as text, then use a fully-featured word processor to produce the cards. To do this, simply save the script to a text file, then open it with your word processor. For an example of a *Trax* layout file designed to facilitate timetable production, see the 'Whitchurch Timetable Model' file in the 'Samples' folder on your CD.

Lever Frame Testing

We saw in chapter 5 how to design a locking table that will prevent conflicting settings of points and signals. How can we be sure that this table is complete – i.e. that there are no combinations of levers allowed by the locking rules which have unforeseen consequences likely to cause problems.

If your lever frame is small, say four or five levers, then you can simply go through all the possible combinations manually. There are 32 possible combinations of five lever settings, and it would not be too much of a task to go through these one by one with *Trax* and check them. If any particular combination is disallowed by the locking rules, *Trax* will indicate this. Similarly, if any combination which is allowed by the locking rules gives rise to a short circuit or two

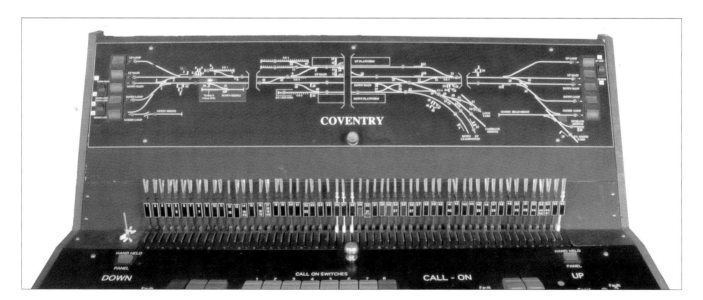

Figure 8.4 *- A manual testing challenge! 56-lever frame built by John Shaw*

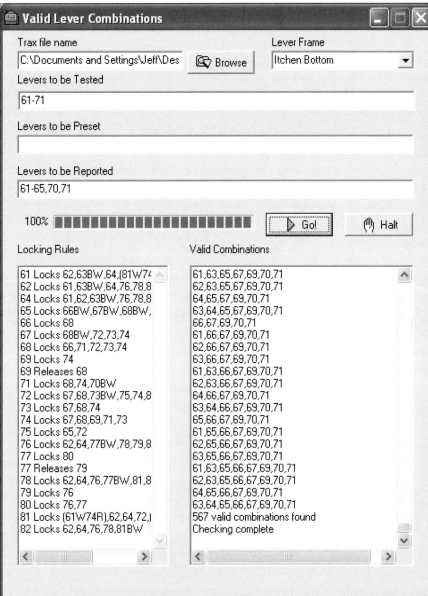

Figure 8.3 - The Frame Test program

However, with the 'Frames' test programme supplied on your CD, things are a made a little easier. Basically, what the programme does is to list only those lever combinations that your locking allows. In addition, it lets you test subsets of the frame. If you have a large frame, it is likely that it will be controlling several areas of the layout relatively independently with little or no interaction between them. You could, for example, have a goods yard operated by one group of levers, and up and down main lines operated by others. You could therefore ask the Frames program to test only those levers related to the goods yard, then subsequently only those related to the main lines. If there were, say, 50 possibilities in each of these groups, there would be a total of 100 to check. Otherwise, you would have to test each of the 50 possibilities in each group with all 50 in the other, giving you 2,500 in total.

To further reduce the number of combinations, you can ask the program to only report those combinations which involve a specific group of levers. This feature is most frequently used in respect of signals. Suppose we have a route which requires points 10,11 and 12 to be set before signal 13 can be pulled 'off'. Clearly, valid combinations would be 10 on its own, then 10 and 11, then 10, 11 and 12 and finally 10,11,12 and 13. However, only the last is of interest, since the others are 'points only' settings which are basically steps on the way to the final route setting. If we tell the program to only report combinations involving signals (which will include 13), then instead of four combinations only one will be reported.

different controllers are connected to the same track, *Trax* will alert you with an error message. You can also check visually that the correct controller is connected to the necessary track sections to complete the signalled movement. You could probably check all 32 combinations in well under an hour.

However, with larger frames the task becomes a great deal more onerous. For example if a frame has 16 levers there are 65,536 possible lever combinations to check. Even if you worked at the rate of checking one of these per minute, it would take you over a thousand hours to check them all! When you get up to 40 levers and above, there are a over a trillion possible combinations, and testing them one by one is not even remotely feasible.

The Frames program, it must be emphasised, is only a simple tool. It will give you a list of valid combinations, that is combinations of lever settings which do not conflict with your locking table. However, it is up to you to test each of them with *Trax*. To test a large lever frame, it may be helpful to break this up into sections. For example, you could test all the levers relating to trains in the 'up' direction, then all those in the 'down' direction. To help with this, you can 'preset' certain levers in the reverse position. The program will then assume these levers are reversed when it processes those you have indicated to be tested. The maximum number of levers you can test in a group is 25.

Using Frames

You can use Frames entirely stand-alone by simply typing or pasting your rules into the 'Locking Rules' edit box. Alternatively, you can load the rules directly from a *Trax* file. Click the 'Browse' button and find your file. The program will locate all lever frames in the layout file. If there are more than one, select which one you want to test from the drop-down list.

Then tell the programme which levers you want to test. You can type individual lever numbers, separated by commas, or you can enter a range, e.g. 5-9, or any combination thereof. Remember that the programme cannot test more than 25 levers at a time. Then, tell the programme what levers, if any, you want to be preset in the 'reverse' position. Finally, in the 'Levers to be Reported' box, indicate what levers you are interested in. For the reasons noted above, these will generally be signal levers. Note that the 'Valid Combinations' will include the point lever settings, but only when they are combined with at least one signal lever.

Finally, click the 'Go' button. Valid lever combinations will be added as they are found and a progress bar will tell you how far through the possibilities the computer has got. If things look like they're going to take forever, click the 'Halt' button and reduce the number of levers you're testing as a group.

Once the program has completed its 'Valid Combinations' list, you can copy and paste this into a text document or similar and start going through it with *Trax* to see that your valid combinations do not include any that should be excluded for any reason. If your locking rules do permit such a combination, then your locking rules are incomplete. Add a new rule to prevent this possibility.

The Lockbars Program

Having arrived at a set of locking rules that allow all valid lever combinations, and exclude all potentially dangerous ones, we must now convert these rules into a physical set of locking bars. In chapter 5, a method of doing this manually was reviewed. For a small lever frame, the manual method is probably perfectly adequate. However, with a larger one the possibility of missing a lock, or inadvertently creating one that we don't want becomes a consideration.

To assist with the creation of a set of locking bars consistent with our locking rules, the 'Lockbars' program was developed. Like the Frame Test program, this is offered 'as is', and it contains only rudimentary facilities. However, it does allow the user to ensure that a design of locking bars covers all the requirements.

As with the Frame Test program, you can either enter a set of rules manually or load them up from a *Trax* layout file. To enter them manually, simply type the rules into the edit box on the left hand side, in the format: 'a Locks b,c,…' or 'a Releases b,c,…'. Once you have entered the rules, click the 'Update Grid' button (the big green arrow) and your rules will be represented graphically on the grid. Instead of typing in your rules, you can, if you wish, fill them in directly on the grid using the 'L', 'BW', 'C' and 'R' buttons.

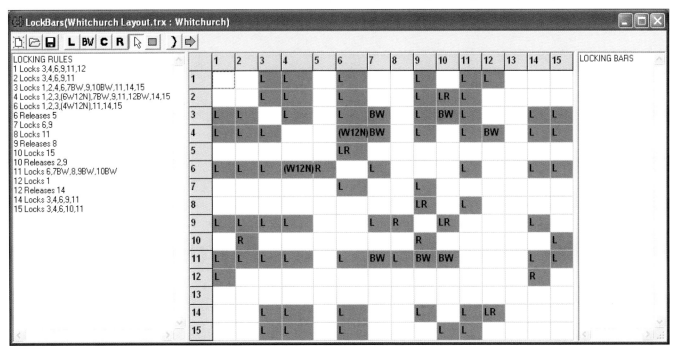

Figure 8.5 - *The Lockbars Program*

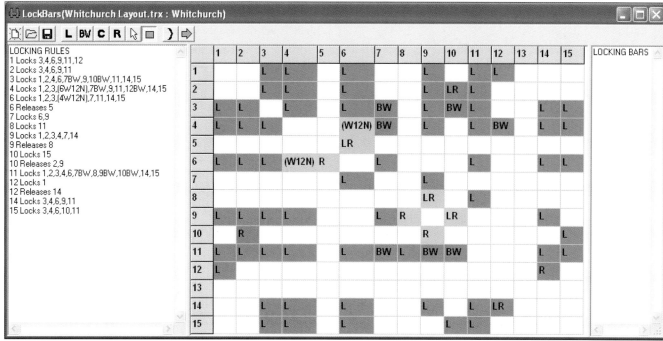

Figure 8.6 - *Initial simplification*

For example with the 'R' (releases) button selected, if you click in the cell for row 3, column 2, you will find that 'R' appears in this cell, and in the cell for row 2 column 3 'LR' appears. These mean, respectively, that 3 releases 2 and that 2 locks 3 in reverse.

Alternatively, if your locking rules have already been included with a lever frame in a *Trax* layout file, use the 'open file' button to access the file. If more than one lever frame is found in the file, you will be asked which one to load up. In figure 8.5, you may recognize the same set of locking rules as we developed in chapter 5. These have been loaded from one of the sample files on your CD. You can save the data in the Lockbars window by clicking 'Save'. This will save it to a 'tlf' or Trax Lever Frame file. Basically, this is a text file in three sections. The heading 'LOCKING RULES' is followed by the locking rules for each lever, e.g. '1 Locks 3,4,6,9'. If a lever releases any others, this is followed by a line, e.g. '10 Releases 2,9'. Then the heading 'LOCK TABLE' is followed by each non-empty cell with row number, column number, type of lock/release, and finally 'S' if the lock is satisfied, all separated by commas. Lastly, the heading 'LOCKING BARS' lists the bars created to satisfy those locks, e.g. 'Lever 12 Locks 1 Releases 14'. You can use this file to save the various steps in the process, so that if you decide you want to go back and try an alternative you can do so.

The initial display (figure 8.5) shows all locks and releases in red, meaning that they have not yet been satisfied, and the 'Locking Bars' edit box on the right is empty – we have not yet created any locking bars. Essentially, what we do is an automated version of the manual process described in chapter 5.

Before we start, however, we do a little preliminary simplification. As was mentioned in chapter 5, locks on adjacent levers are most easily done by soldering a bar on the lever which is doing the releasing such that it prevents the adjacent lever from moving until it is reversed. This can be applied to the rules '6 Releases 5', '9 Releases 8', and '10 Releases 9'. Therefore, mark these rules as 'satisfied' by changing their related squares from red to green. Click the green square on the toolbar, then click row 6 column 5, row 9 column 8 and row 10 column 9. Next, we consider the conditional lock. It is most unlikely that conditional locks can be easily accommodated on the same locking bars as normal locks. Therefore, we again mark these green manually (in this case, by clicking row 4 column 6). The program will not then try to combine conditional locks with other locks and releases. As a result of this, your screen should now look like figure 8.6. We can now proceed with the process of combining individual locks into locking bars.

As with the manual method, we are looking for patterns – basically, rows that have a lot in common. We first of all notice that rows 1, 2 and 14 are identical except for a couple of cells. We also notice that row 15 is very similar. In fact, we can make it more so by noticing that we have no lock between 15 and 9. However we do have a lock between 15 and 10, and since 10 releases 9 we can add a lock between 15 and 9. This is technically superfluous, but it does increase the similarity between rows 1, 2, 14 and 15. Click the 'L' button on the Toolbar, then click in the cell in row

15, column 9. You will notice the program updates the locking rules automatically.

Next, we combine the rows. With the 'Select a Row' tool (the little arrow) we therefore click any cell in row 1 – notice that the whole row now turns pink (note – you must click within the row itself, not on the grey 'row number' cell). Next click any cell in row 2 – this too turns pink. Repeat for rows 14 and 15. Then click the 'Combine Rows' tool (the curly bracket). All the *common* locks in rows 1 and 2 will now turn green, meaning they have been satisfied, and our first locking bar will appear on the right. This will be 'Levers 1,2,14,15 Lock 3,4,6,9,11'.

Proceed in the same manner until all of your required locks have been met. We notice that 3 and 4 are very similar, and can be made more so by adding '3 Locks 12BW' (no operational restrictions are caused by this since no move involving 12 can be made whilst 3 is pulled anyway). Combining rows 3 and 4 will then give you a locking bar 'Levers 3,4 Lock 7BW,9,11,12BW'. There is not a lot more that the remaining rows have in common, so next click the row with the largest number of unsatisfied locks. This will be row 11, click this and 'combine' it with itself (just click the curly

bracket). This will give you a locking bar : 'Lever 11 Locks 6,7BW,8,9BW,10BW'. Do the same thing for rows 3, 7, 10 and 12 and you will end up with a set of locking bars as follows:

Levers 1,2,14,15 Lock 3,4,6,9,11
Levers 3,4 Lock 7BW,9,11,12BW
Lever 11 Locks 6,7BW,8,9BW,10BW
Lever 3 Locks 4,6,10BW
Lever 7 Locks 6,9
Lever 10 Locks 15 Releases 2
Lever 12 Locks 1 Releases 14

Notice that these are very similar to those we obtained manually in chapter 5. There is no single 'correct' set of locking bars. There are several ways to achieve the required lockings and so long as they use no more bars they are all equally good.

Finally, we would need to add to these locking bars any additional ones to cover the rules we excluded from the automated process, i.e. conditional or sequential locks. Having completed the set of locking bars required, we go on to construct them as described in chapter 5.

Signals at Basingstoke

BOOKS

The following is a list of books on Railway Signalling and associated topics, which may be found useful if you want to research specific prototypes or investigate in more detail the topics touched on in the first two chapters of this book. It is not a comprehensive list; it contains only a selection of the many books on this topic. Unfortunately, some are out of print, but you may find a second-hand copy at one of the railway booksellers. Alternatively, your local library might have access to a copy which they can find for you. If you want to look in your library for railway signalling books, the Dewey classification number is 625.165.

Two Centuries of Railway Signalling
Geoffrey Kitchenside & Alan Williams.
2nd ed. OPC, 2008.
ISBN: 9780860936183

A Pictorial Survey Of Railway Signalling
D H Allen and C J Woolstenholmes.
Haynes, 1991.
ISBN: 9780860934530

BR Signalling in Colour for the Modeller & Historian
Robert Hendry.
Midland Publishing, 2001
ISBN 9781857801149

BR Signalling Development in Colour for the Modeller and Historian
Robert Hendry.
Ian Allan, 2009.
ISBN: 9780711033627

Signalling in the Age of Steam
Michael A. Vanns
Ian Allan, 1995. Reprinted 2005.
ISBN 0711026238

An Illustrated History of Great Northern Railway signalling
Michael A. Vanns
OPC, 2000.
ISBN: 9780860935452

A Pictorial Record of LNER Constituent Signalling
A. A. Maclean
OPC, 1983
SBN 0860931463

A Pictorial Record of LNWR Signalling
R.D. Foster
OPC, 1982
SBN 0860931471

A Pictorial Survey of London-Midland Signalling
D H Allen & C J Woolstenholmes
OPC, 1996
ISBN 9780860935230

A Pictorial Record of LMS Signals
L G Warburton & V R Anderson
OPC, 1972.
SBN 0902888099
Paperback reprint published by Kevin Robertson under the Noodle Books imprint, 2010
ISBN 9781906419417

A Pictorial Record of Great Western Signalling
By A. Vaughan
OPC, 1973
SBN 0902888080

A Pictorial Record of Southern Signals
G A Pryer.
OPC, 1977.
SBN 902888 81 1

WEBSITES

These days, the World Wide Web is becoming as useful a source of information as books. The following are websites specialising in various aspects of railway signalling and lever frames. Again, the list is not claimed to be comprehensive and, sadly, it may become out of date in time. Just as books go out of print, websites sometimes become defunct. To get an up-to-date list, you can try typing 'railway signalling websites' into your favourite search engine, but be warned: at the time of writing, doing this produced just over three million results! Here are a few of them:

www.signalbox.org
John Hinson's website describes the principles behind railway signalling in Great Britain, but some coverage of signalling around the world will also be found. There is also a comprehensive list of other signalling websites and links to places where you can go and see signalling equipment.

www.railsigns.co.uk
This website is nothing if not ambitious! The aim is to illustrate every signal indication or lineside sign that exists today

or has ever existed in the past and to explain its meaning.

www.wbsframe.mste.co.uk
Mark Adlington's website provides information on signal boxes that used Westinghouse miniature power lever frames. British Railways, London Transport, and their predecessors all used these lever frames, as well as overseas railway companies from 1910 - 1960.

www.s-r-s.org.uk
The Signalling Record Society maintains and shares knowledge of railway signalling and operation in the British isles and overseas. They possess much archive material and publish books, photographs and drawings which may be purchased.

www.svrsig.org
This site describes signalling on the Severn Valley Railway and the activities of the volunteer S&T Department. Mechanical locking and some of the principles of Signal Engineering are explained.

www.distantwriting.co.uk
Although Steven Roberts' website is primarily a history of the telegraph companies in Britain between 1838 and 1868, it contains an interesting section on railway telegraphy.

www.lymmobservatory.net
John Saxton's website (as you might guess from its URL) is principally astronomical, however it does contain a very interesting section on railway signalling including a block post simulator, signal box diagrams, train register extracts etc.

www.roscalen.com
Another web site with a wide range of topics. However, Adrian the Rock's contains an extensive collection of photographs of signals and signal boxes etc.

www.tillyweb.biz
John Tilly's website is primarily concerned with railway and other photography, level crossings, railway signal engineering, transport ticketing and associated areas of interest.

www.simsig.co.uk
This is a fun site if you like playing computer games. SimSig places you in the signaller's seat and lets you control the trains. It recreates the environment of a signalling control centre and it is up to you to route the trains to their destination and do your best to keep them on time.

SIGNALLING PARTS

The following suppliers make parts which will help you you to build your own signals and lever frame.

www.modelsignals.com
Model Signal Engineering (Andrew Hartstone) supplies a wide range of semaphore signalling parts for British railway companies, including pre-grouping companies and private contractors such as Stevens & Co, Saxby & Farmer etc. All scales are covered from 2mm/ft up to 16mm/ft. MSE also produce a seven lever frame, which can be ganged together to make frames of 14, 21, 28 levers etc.

Scale Signal Supply
This company, of 135 Green Meadows, Bolton, BL5 2BW, do not, at the time of writing, have a website. However you can see the full range of their signal kits on Invertrain's website (**www.invertrain.com**). Scale Signal Supply also produce a kit of parts to make a 1/12th scale Stevens & Co lever frame – these parts were used to make the frame featured in chapter 5.

www.modratec.com
Modratec, an Australian company, produce a lever frame with mechanical interlocking. A simple software program enables you to produce a schematic version of you layout's track plan and design locking for it.

www.borg-rail.co.uk
Servo control boards for signals and points,semaphore and colour light Signals.

www.winterleyproducts.co.uk
John Shaw can design and build an interlocking frame to suit your track plan. For examples of his work, see figures 1.5 and 4.9. The frames are built from Scale Signal Supply parts.

APPENDIX 2

BELL CODES

Bell codes differed from railway to railway, and over time. The following is a list of the common bell codes used on BR in the 1960s. For more extensive lists, see the sources listed in Appendix 1. The notation 3-1 etc. means three rings of the bell in quick succession, followed by a short pause, then a single ring. Where a signalbox had to communicate with more than one other box, the bells would be designed to sound different, e.g. one would be a higher-pitched bell, and one a deeper-pitched gong.

The following bell codes are used to describe the train awaiting entry to the next section and to ask for 'line clear'. If the line is clear, the signal box in advance will repeat the bell code back. If not, the line is blocked, and the box in rear must wait then re-request.

There were also a number of local codes in use between specific signal boxes at larger stations and junction.

Bell Code	Train Description	Train Class
4	Is line clear for express passenger trains, newspaper train, breakdown van train or snow plough going to clear the line, light engine going to assist disabled train or officer's special train not requiring to stop in section?	A
3-1	Is line clear for ordinary passenger train, mixed train, or breakdown van train not going to clear the line?	B
1-3	Is line clear for branch passenger train? Where authorised.	B
1-3-1	Is line clear for parcels, fish, fruit, horse, livestock, meat ,milk, pigeon or perishable train composed entirely of vehicles conforming to coaching stock requirements?	C
3-1-1	Is line clear for express freight, livestock, perishable or ballast train, pipe-fitted throughout with the automatic brake operative on not less than half of the vehicles?	C
2-2-1	Is line clear for empty coaching stock train (not specially authorised to carry "A" headcode)?	C
5	Is line clear for express freight, livestock, perishable or ballast train, partly fitted, with the automatic brake operative on not less than one third of the vehicles?	D
1-2-2	Is line clear for express freight, livestock, perishable or ballast train, partly fitted, with the not less than four braked vehicles next to the engine and connected by the automatic brake pipe, or express freight, livestock, perishable or ballast train, with a limited load of vehicles not fitted with the automatic brake?	E
3-2	Is line clear for express freight, livestock, perishable or ballast train, not fitted with the automatic brake?	F
1-1-3	Is line clear for engine with not more than two brake vans?	G
2-3	Is line clear for light engine or light engines coupled?	G
1-4	Is line clear for through freight or ballast train not running under class "C", "D", "E" or "F" headcode?	H
4-1	Is line clear for mineral or empty wagon train?	J
1-2	Is line clear for branch freight train? Where authorised.	K
3	Is line clear for freight, mineral or ballast train, stopping at intermediate stations?	K

BELL CODES

The class of train is indicated by headlamps on the front of the engine, as in the diagram.

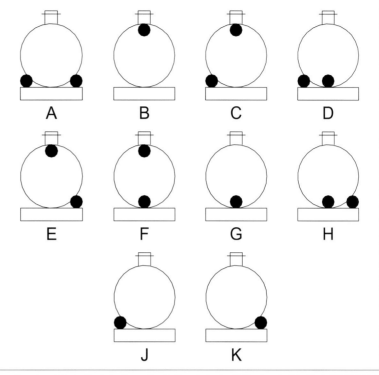

The bell codes below are used to inform other signal boxes about train movements, and for general information. For example, when a 'shunt-ahead' move is to be made, the signal box in advance must be sent '3-3-2 Shunting into forward section', as described in chapter 3.

1	Call attention
2	Train entering section.
3-5-5	Warning acceptance
2-1	Train out of section, or obstruction removed.
2-4	Blocking back inside home signal
3-3	Blocking back outside home signal.
3-5	Train cancelled
3-2-3	Train drawn back clear of section.
6	Obstruction danger.
6-2	Train an unusually long time in section.
7	Stop and examine train
9	Train passed without tail lamp, to box in advance.
4-5	Train passed without tail lamp, to box in rear.
5-5	Train divided
1-5-5	Shunt train for following train to pass.
2-5-5	Train or vehicles running away in wrong direction.
4-5-5	Train running away in right direction.
16	Testing block indicators and bells.
3-3-2	Shunting into forward section.
8	Shunt withdrawn.
2-3-3	Working in wrong direction.
5-2	Train clear of section.
2-5	Train withdrawn.
5-3	Last train incorrectly described
5-5-5	Opening of signal box.
5-5-7	Closing of signal box where section signal is locked by the block.
7-5-5	Closing of signal box.

APPENDIX 3

WHITCHURCH TOWN ENIGMA

The following pages give the complete circuit diagram for the Whitchurch Town 'Enigma' box. The function of the box, as described in chapter 4, is to take the inputs from the lever frame and to use these to operate signals and points then connect controllers to appropriate track sections. Inputs to the box from the lever frame are called L1, L2 etc. These are normally at -12v, but if a lever is pulled, a microswitch operates to connect the input to +12v. Therefore, levers 6, 9, 10 and 12 can operate point motors directly. However, signals are operated by solenoids, so diodes D01, D02 etc. are required to stop the solenoid operating on -12v. Notice that levers 3 and 4 each operate one of two signals depending on the setting of points 10 and 12. Relays operated by levers are called RL10, RL12 etc. Again, diodes are necessary since relays can operate just as well on -12v as +12v. Notice that relays RL1 and RL15 are designed to operate on -12v. When a fiddle yard operator pushes a 'Train Ready' button, the appropriate relay comes on and stays on until the train is signalled into the station. The relay operates a flashing LED on the lever frame panel to alert the signalman that a train is waiting.

The outputs labelled F1R, F1G etc. denote 'feed 1 connected to red controller', 'Feed 1 to green' etc. These illuminate LEDs on the track diagram to show which controller is powering which sections of track, and also control the relays that make these connections. The connections we require to make are those shown in table 4.1 (chapter 4). Note that the 'N' or Newbury controller is colour-coded red, the 'S' controller yellow and the 'Y' controller green. The right hand side of the first page of circuit diagram is concerned with the logic of this process. Notice, for example, that pulling lever 4 will connect feed 3 to green *unless* lever 7 is pulled, in which case it has the same effect as lever 14.

The reason for this is explained in chapter 4.

The second part of the circuit diagram is concerned with the track relays. These are called RT..., for example, RT1R connects feed F1 to the 'red' controller. Notice that the track relays controlling the sections at the outer ends of the layout (feeds F1 and F3) are 'latched'. That is, they stay on even after the starter signal (S2 or S14) is put back, so long as the route remains set. This allows for a more realistic operation, in that the signal can be put back to danger as soon as the train has passed it, rather than waiting until the train has come to a halt at its destination. C1, C2 and C3 are the 'live' outputs of the red, green and yellow controllers respectively.

The final section of the circuit diagram is concerned with switching the crossing vees of the various points and with releasing the ground frame. Many modellers choose to do this under the layout using the switch terminals available on most point motors. However, I prefer to use relays to do this. The location of crossing vees and track feeds can be found in the Trax file 'Whitchurch Track Sections and Feeds' in the 'Samples' folder on your CD. The file 'Whitchurch Relay Logic' illustrates all of the relays on this diagram, along with the lever frame and track plan. You can use this to check out how the relay logic works. The ground frame is released by lever 7 on the main frame, and opened by 'king' lever 1 on the ground frame itself. If the king lever is pulled without lever 7 a buzzer sounds. If the signalman wants the ground frame to close (e.g. to back a train into the yard) he must press 'ground frame request'. This will set up a flashing LED on the ground frame, until such time as the frame is closed (lever K1 returned to normal).

Whitchurch Town: 12 inches to 1 foot scale.

WHITCHURCH TOWN ENIGMA

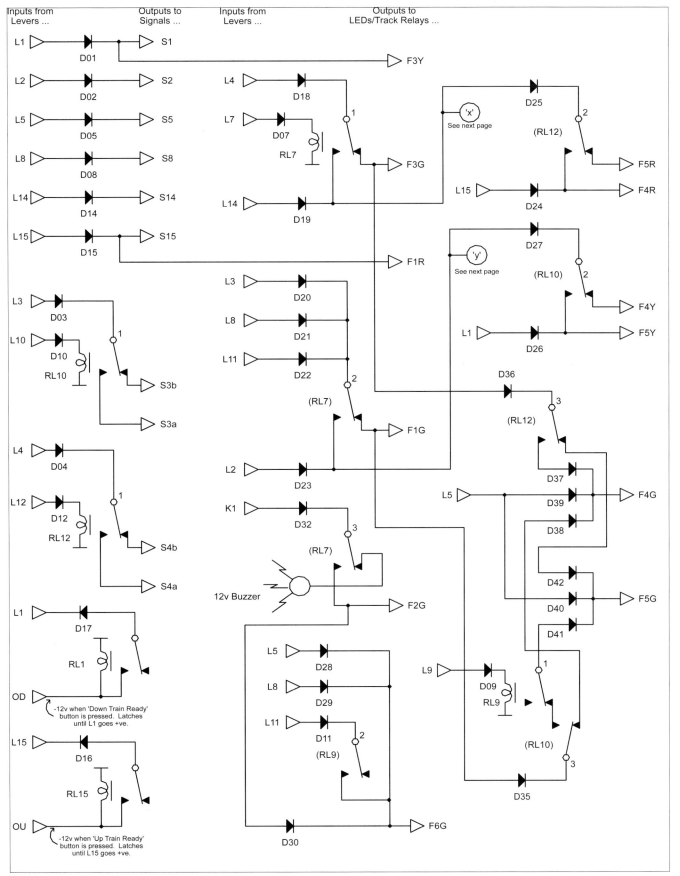

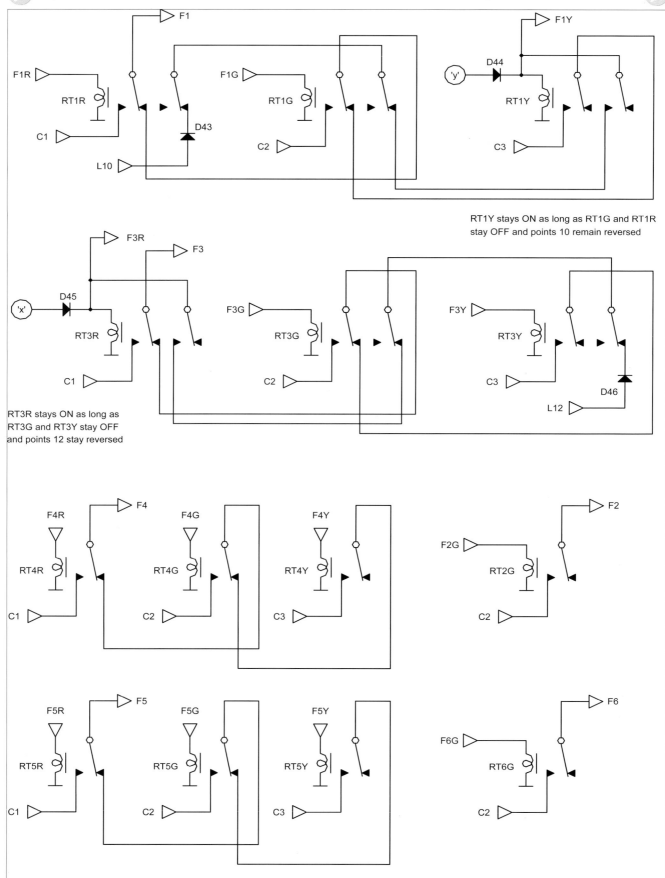

RT1Y stays ON as long as RT1G and RT1R stay OFF and points 10 remain reversed

RT3R stays ON as long as RT3G and RT3Y stay OFF and points 12 stay reversed

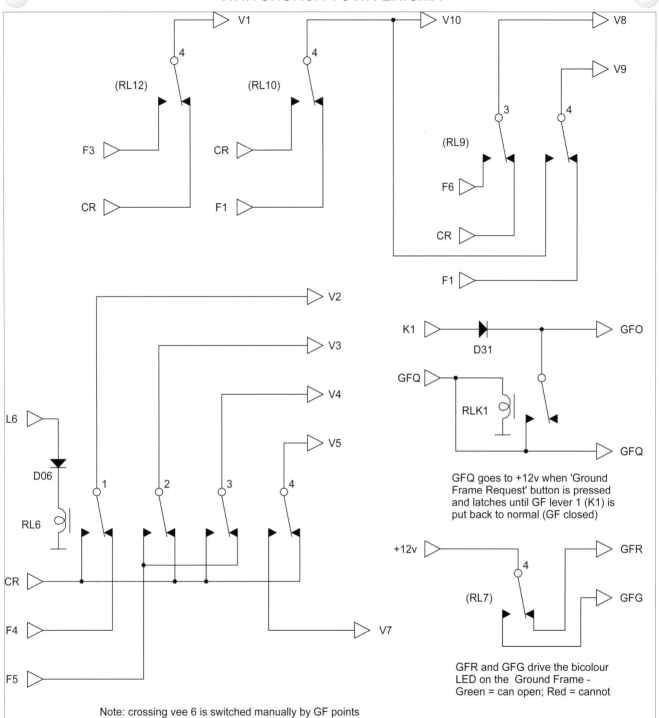

GFQ goes to +12v when 'Ground Frame Request' button is pressed and latches until GF lever 1 (K1) is put back to normal (GF closed)

GFR and GFG drive the bicolour LED on the Ground Frame - Green = can open; Red = cannot

Note: crossing vee 6 is switched manually by GF points

APPENDIX 4

ITCHEN BOTTOM ENIGMA

There follows the full circuit required to implement the complete switching logic involved in the 'Itchen Bottom' layout described in the last section of chapter 4. As is mentioned in the text, the lever frame was already built with 22 levers. A good deal of the complexity is concerned with making just 22 levers drive 34 signals and 8 sets of points.

The lever frame is slightly different from that of 'Whitchurch' in that the levers operate simple on-off switches, i.e. when the lever is normal there is no output voltage, when it is reversed there is +12v. This means that to get a polarity-changing output for driving point motors we require a pair of changeover contacts on each of the 'points' relays. The points operate on ±9v, which gives a slightly more realistic speed of operation.

The first group of relays in the circuit diagram below are all concerned with selecting a particular signal when a particular lever is pulled. This logic is contained in table 4.4. Note that only four levers operate just a single signal (70, 79, 80 and 82). For the rest, the signal lever pulls are combined with the route set by the point levers to determine which signal comes 'off'. In the case of signal lever 71, this is further complicated by the fact that if lever 70 is pulled as well, a different controller is to be selected.

Having resolved all the signal logic, we now need to determine from the signals which controllers are connected to which track sections. This is most easily done by means of a diode matrix, as shown. This is a physical realization of table 4.5. For example, we see from table 4.5 that track

section Up2 is to be connected to the blue (B) controller is signal 82, 81a, 81b, 81c or 81d is off. Thus we take a diode from each of these to the right hand vertical wire in the matrix labelled Up2B. Applying similar logic to each track section/controller combination we end up with the complete matrix as shown.

The final job is to use the outputs from the diode matrix to power the final set of relays, which are our track relays. These are called, for example, RTUp2B – the track relay which connects Up2 to the 'B' controller. There are a lot of these because we have 14 track sections and 4 controllers. Not all 56 possible combinations are required, but we still need a good number of them.

This circuit is a good deal more complicated than that for Whitchurch. However, the basic principles are exactly the same. We use logic relays to determine which combination of levers elicits which track/controller connections, then we use the latter to drive a set of track relays to actually make those connections. If the logic is not very extensive then you can dispense with the diode matrix and just incorporate what few diodes are needed in among the logic relays. If, as here, the logic is particularly convoluted, then a diode matrix is a neater solution. In the case of Itchen Bottom, it was decided to use 'Tortoise' point motors and to use these to switch crossing vee polarity, so that the equivalent of the final section of the 'Whitchurch' circuit is not necessary. A Trax version of the Itchen Bottom layout can be found in the samples folder on your CD.

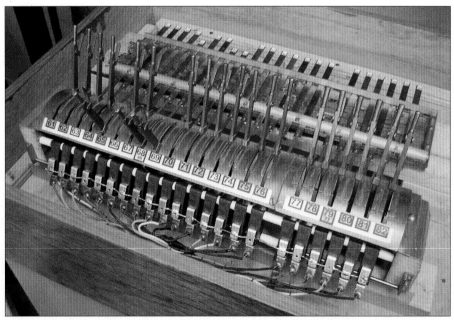

The scratch-built lever frame driving the relays described in this appendix. Note the locking tray behind the levers and the row of microswitches in front.

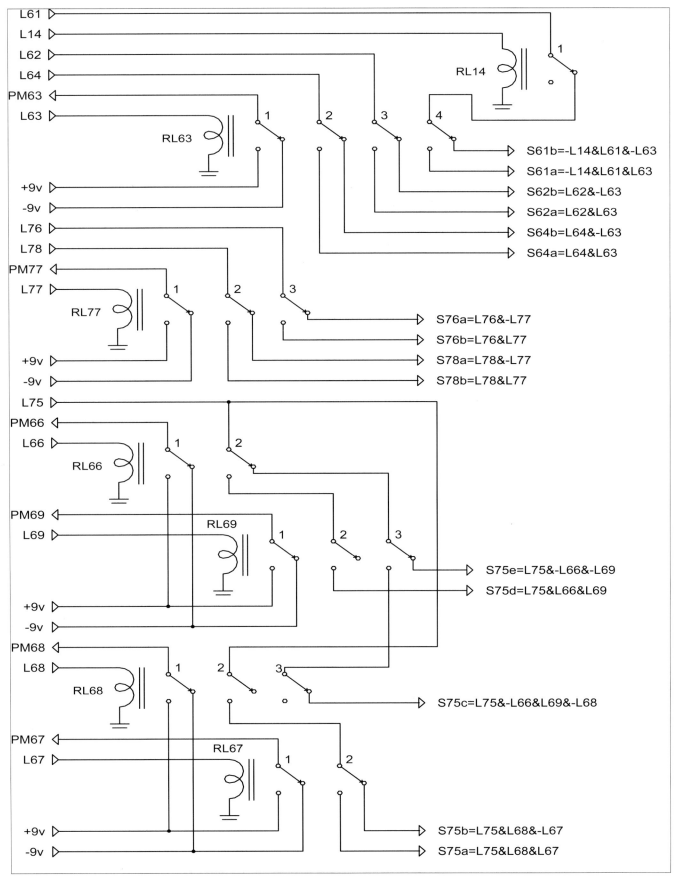

L61
L14
L62
L64
PM63
L63

RL14

RL63

1

2

3

4

1

S61b=-L14&L61&-L63
S61a=-L14&L61&L63

+9v
-9v
L76
L78
PM77
L77

RL77

1

2

3

S62b=L62&-L63
S62a=L62&L63
S64b=L64&-L63
S64a=L64&L63

S76a=L76&-L77
S76b=L76&L77
S78a=L78&-L77
S78b=L78&L77

+9v
-9v
L75
PM66
L66

RL66

1

2

PM69
L69

RL69

1

2

3

S75e=L75&-L66&-L69
S75d=L75&L66&L69

+9v
-9v
PM68
L68

RL68

1

2

3

S75c=L75&-L66&L69&-L68

PM67
L67

RL67

1

2

+9v
-9v

S75b=L75&L68&-L67
S75a=L75&L68&L67

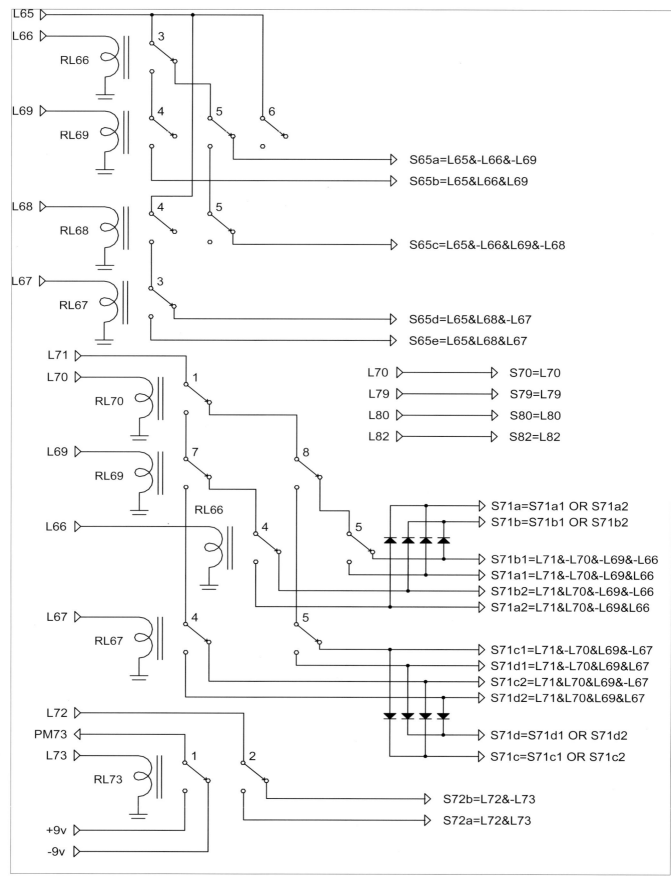

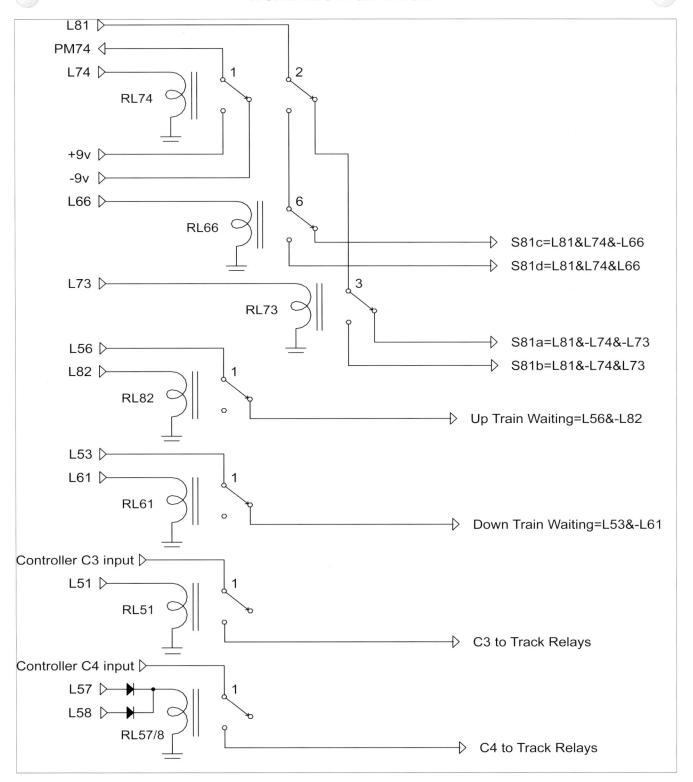

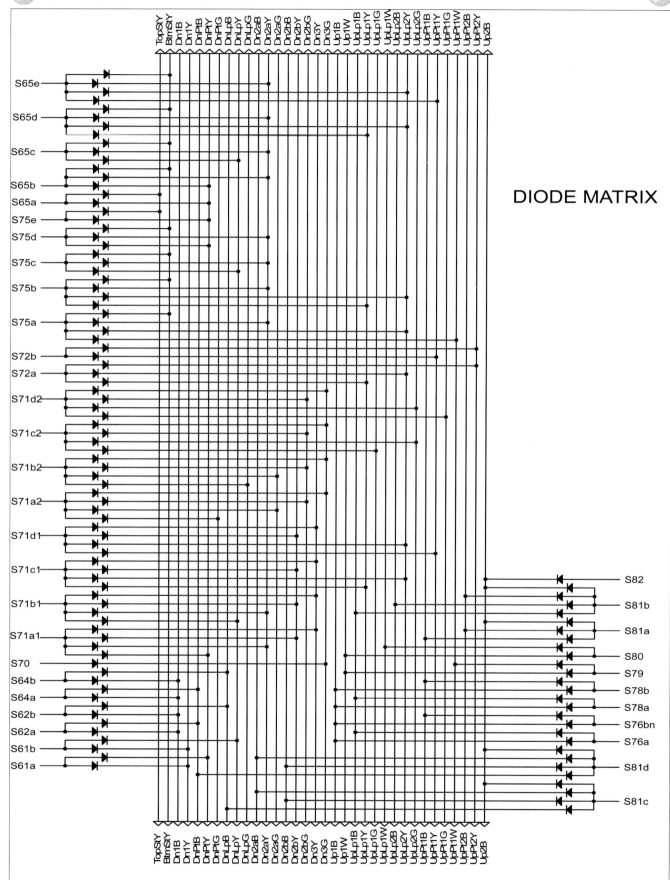

DIODE MATRIX

ITCHEN BOTTOM ENIGMA

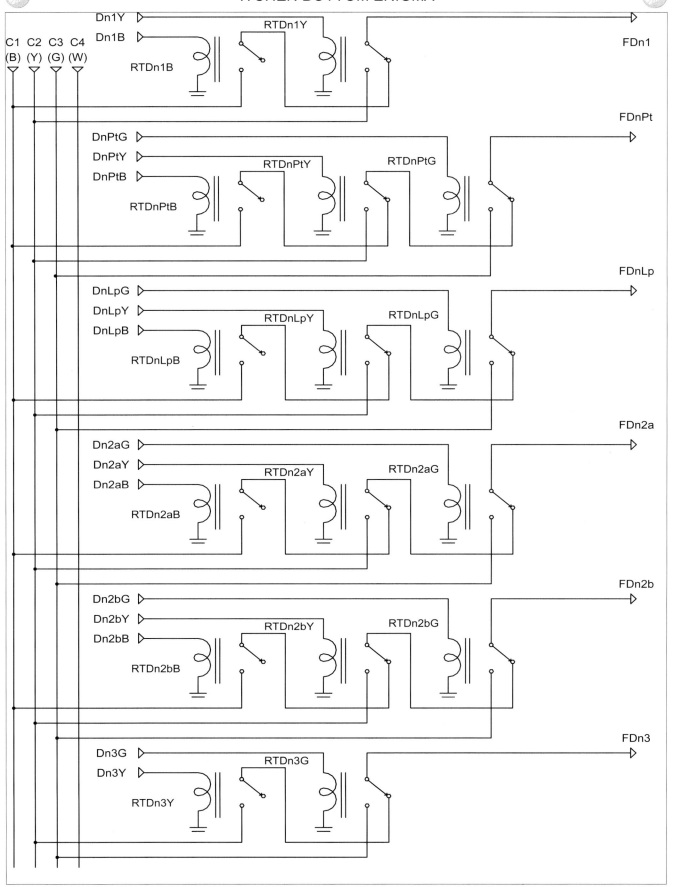

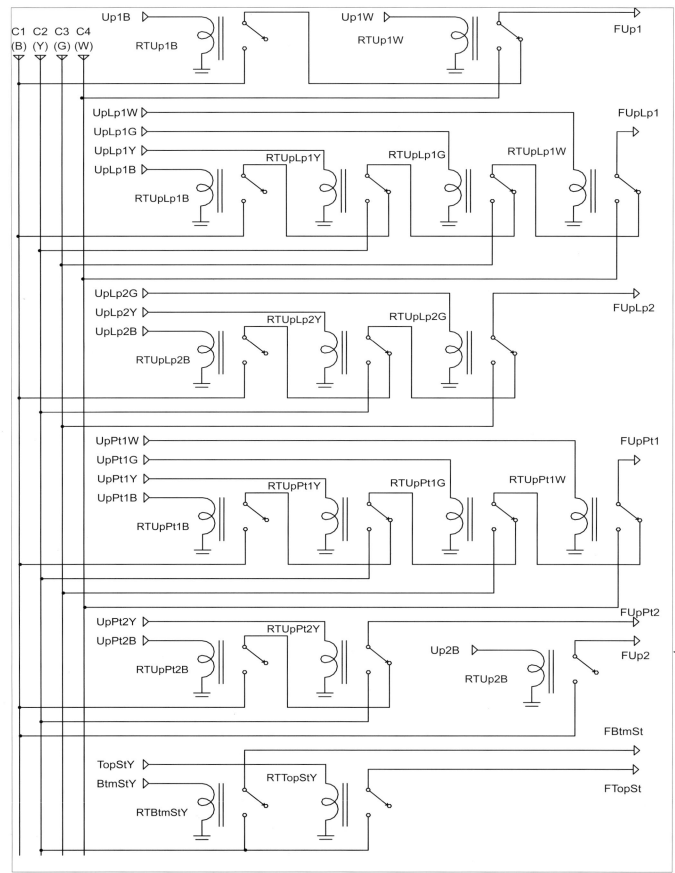

COLOUR PLATES

Plate I – This splendid gantry is at Burnden Junction on the Lancashire and Yorkshire route from Manchester Victoria to Bolton (you can see the rather fine clock tower on Bolton Town Hall in the background). A down through train is signalled, whilst waiting for the up starter is class 47 number 47537 Sîr Gwynedd-County of Gwynedd hauling a class 87 electric locomotive, number 87002 Royal Sovereign and the 09:10 Glasgow to Poole train through this non-electrified section on 23rd September 1984. Photograph by David Ingham.

Plate II - Unusual these days to see 'new' semaphore signals, but these at Yeovil Pen Mill were renewed in 2009. Geof Sheppard's picture shows refurbished ex-LMS upper quadrant arms on tubular posts surmounted by GWR finials! The signals control movements from each of the two platforms towards Weymouth (left) and Yeovil Town (right).

Midland Railway (pre-1911)

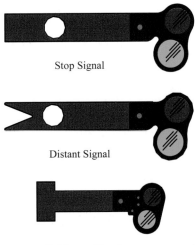

Stop Signal

Distant Signal

Calling-on Signal

London & North Western Railway

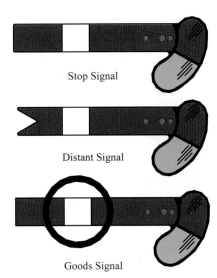

Stop Signal

Distant Signal

Goods Signal

Note: After 1911, the white spot was replaced by a white vertical stripe. Rear of signals was white with a black spot, although in some cases the distant signal had a central black band along the full length of the arm instead of a spot.

Great Western Railway

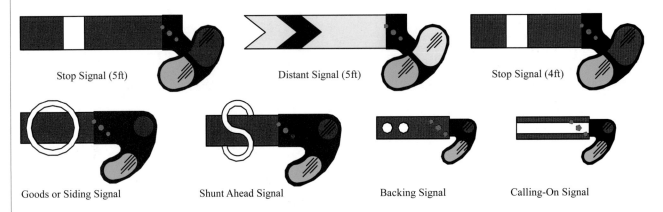

Stop Signal (5ft)

Distant Signal (5ft)

Stop Signal (4ft)

Goods or Siding Signal

Shunt Ahead Signal

Backing Signal

Calling-On Signal

Notes : The 5ft arms were used where the signal was mounted on a post of 26ft or more for extre visibility. The two circles in the backing signal are actual holes, not painted spots like the M.R. signals!

Upper Quadrant Arms (LMS, LNER, SR and early BR)

Stop Signal

Distant Signal

Subsidiary Signal...   raised to show 'C', 'S' or 'W'

Notes : Unlike the above, these signals worked in the upper quadrant and had the signal lamp on the left hand side of the post (as viewed from the front). The lamps glowed yellow, so that when this shone through the blue spectacle glass, the light was perceived as green.

Colour Light Signals

Three Aspect Signals

Danger - Stop   Caution   Clear

Standard three-aspect colour lights, as recommended in A. F. Bound's 1924 report, and later adopted by all railway companies except the Great Western. The earliest colour-light signals had only two aspects: red and green.

Four Aspect Signals

Danger - Stop   Caution   Preliminary Caution   Clear

Where high speeds are involved, a second yellow light is used to indicate 'preliminary caution'. The single yellow means that the next block section is clear, but be prepared to stop at the one after. Double yellow means the next two sections are clear, but be prepared to stop after these. On lines where speeds over 125mph are permitted, a fifth aspect, flashing green, is used to indicate that the next light is green also.

GWR Colour Light Signals

Danger - Stop   Caution   Clear

GWR colour lights employed moving spectacles so that a single lamp could display different colours exactly like semaphores seen at night. The above shows the equivalent of stop and distant signals mounted one above the other on a single post.

Direction Indicators

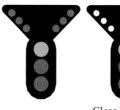  

Clear straight on   Clear Diverging to left   Clear diverging to right

Three-aspect signals with early S.R. pattern three-light direction indicators.

Three-aspect signal with later-style five lamp direction indicators. This one signals the second from left of six diverging routes

Subsidiary Signals

On   Off   On   Off

Early subsidiary colour light signals were miniature versions of the full-size two-aspect signals. The 'yellow' signal could be passed at danger except for the route to which it applied.

On   Off   On   Off

Early position light signals had three white lights - two horizontal lights for 'on' and two lights inclines at 45° for 'off'. Later, one of the white lights was substituted for red.

Left - Plate III *– A small selection of the many and varied types of semaphore signal arm. See Appendix 1 for books and web-sites giving details of your favourite railway's signals.*

Above - Plate IV *– Colour-light signalling was first introduced on the Liverpool Overhead Railway in 1920, and gradually replaced semaphore signalling over the entire system, with the exception of a few outposts, principally in the Western Region (where else!).*

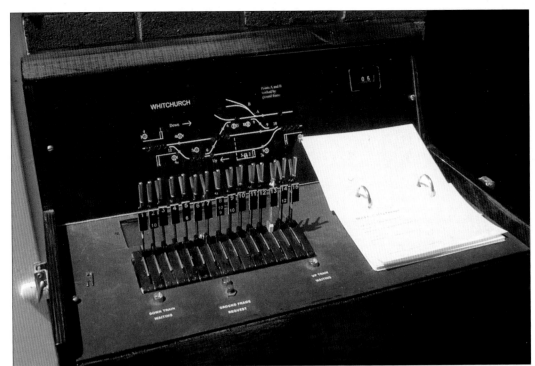

Plate V – *The lever frame whose construction is described in chapter 5, complete with its track diagram and timetable 'cue cards'. The flashing LEDs below levers 1 and 15 indicate that a down or up train is waiting to enter the section. The blue push-button sets off a flashing LED to indicate to the goods yard operator that a train is waiting to enter the yard.*

Plate VI – *A lever frame built by John Shaw for David Deere-Jones' magnificent model of Coventry station. The model is of the section between Warwick Road and Quinton Road bridges. Three signal boxes on the prototype controlled this area so they were amalgamated and a new frame and interlocking designed around the area modelled, including the approaches from the London and Birmingham ends. Signalling details known at the time did not include shunt signals to allow the various moves within station limits to be made, so the green switches act as call-on switches to allow moves normally made when authorized by the signalman by green flag or lamp. Whilst a call-on switch is activated, the signals protecting those track sections are electrically locked normal. These switches select the call-on controller allowing the signalman to make the shunt move.*